Garden Way's

Red & Green
Tomato Cookbook

Janet Ballantyne

Garden Way Publishing
Pownal, Vermont 05261

Acknowledgments

I'd like to thank all the people at Garden Way's Meeting House who sampled the recipes in this book and extend a special thanks to Andrea Chesman for her continued support and enthusiasm.

Cover design by Trezzo/Braren Studios
Cover photo by Joan Knight
Photographs by Erik Borg
Illustrations by Elayne Sears
Copyright © 1982 by Storey Communications, Inc.

The name Garden Way Publishing is licensed to Storey Communications, Inc.
by Garden Way, Inc.

Printed in the United States by Capital City Press.
Cover printed by Federated Lithographers.
Fourth printing, May 1984

Library of Congress Cataloging in Publication Data
Ballantyne, Janet, 1952-
   Garden Way's Red & Green Tomato Cookbook.

   Includes index.
   1. Cookery (Tomatoes)   I. Garden Way Publishing.
II. Title.   III. Title: Garden Way's red and green
tomato cookbook.
TX803.T6B34   641.5′6′5642   82-2884
ISBN 0-88266-262-7   AACR2

Contents

Introduction

Picture yourself in a summer garden with delicious tomatoes hanging in clusters of red and green, promising succulent delight. Pluck one of those ripe tomatoes, smooth and silky to the touch, and let the sweet warm flavor of summer explode in your mouth. This is the first way to enjoy a tomato.

There are countless ways to appreciate the rich, juicy goodness of tomatoes. This is a book about enjoying the taste you associate with joys of summer—all year long.

The tomato is the most popularly grown vegetable in America for a number of reasons. First, it's one of the tastiest vegetables around. Also, it's a very nutritious food, ranking high in vitamins A and C and potassium. As you can see from all the recipes collected here, the tomato is a versatile food. And it is easy to grow.

Even if you have never gardened before, you can expect success with tomatoes. They can be grown in every part of the country—in large fields or gardens as small as a pot on a balcony.

Growing your own tomatoes enables you to choose your varieties. As a gardener you have many more types of tomatoes available to you than shoppers at a supermarket (who rarely get a vine-ripened tomato in any case).

Italian plum tomatoes are especially good for sauces and purees because they have more pulp and less water and seeds than other tomatoes. Cherry tomatoes are great in salads, on vegetable platters, and as stuffed appetizers. And no garden is complete without an all-round, sweet, juicy, slicing tomato, terrific for eating fresh right out of the garden or combined with other flavors in a cooked dish.

Choosing the best varieties to grow depends on the length of your growing season, your climate, and what you like in a tomato. The varieties you choose should ripen at different times so you can stretch the harvest season. Also, by planting a few different varieties you reduce the risk of losing your entire crop to disease.

Here in Vermont, Garden Way's Master Gardener Dick Raymond usually recommends that gardeners raise Pixie, Early Girl (both early season), Better Boy, Big Girl, and Rutgers (main crop varieties). Burpee has a real late season variety

called the Long-Keeper. According to the Burpee people, these light-colored, orange-red tomatoes will stay fresh for 6–12 weeks, if harvested when ripe or partially ripe and stored properly. Although the flavor of these tomatoes can't compare to a vine-ripened Big Boy, for example, it will beat the taste of the supermarket product. Also, early season tomatoes don't seem to have as much flavor as the tomatoes that ripen later in the season. So plant just enough of the early tomatoes to get you started, and leave plenty of space for the other varieties.

Your neighbors, or your local County Extension Service agent will be the best sources of information on specific tomato varieties that will grow well in your area.

To determine how many tomato plants to grow, figure that each tomato plant will produce about a half bushel, or 30 pounds of tomatoes. It takes about 3 pounds of fresh tomatoes to yield a quart of canned tomatoes. So that's about 10 quarts of tomatoes per plant. For eating fresh and preserving enough tomatoes for a small family, 10–18 tomato plants should be enough.

Tomatoes transplant well, so it is a good idea to start the plants indoors about 6–8 weeks before the average last killing frost date.

For starting your tomato plants, choose a container that has holes in the bottom for drainage. Commercial seed flats work well, but you can use any sort of box, pot, or paper cup, as long as there are drainage holes. Fill your container with a sterile potting mix or soil formula. Sprinkle the seeds about ½ inch apart on top of the soil. Pat the seeds down into the soil and cover with a thin layer of soil. Enclose the entire container with a plastic bag to keep in the moisture. Keep the trays in a warm sunny place. In a few days the seedlings will appear. Remove the plastic bag. Do not allow the soil to dry out.

When the plants are 3 inches tall, transplant the seedlings into a deeper flat or pot to give the plants more room to grow. Remove all but the uppermost leaves of each plant, and bury the plants in soil until only the top leaves show. The tiny "hairs" on the tomato stem will create a strong root system for the plant, and the more roots the plant has, the sturdier and healthier it will be.

When the tomatoes are about 10 inches tall, you can transplant them a second time, as Dick Raymond does. Half-gallon milk cartons are ideal for this job. Place the plant in the carton. Pluck off all but the top leaves again, and put soil around the plant so only these top leaves show.

Ten days before you are ready to put your

plants in the garden (after the last average frost date), begin to harden off the plants to prepare them for their life in the great outdoors. On the first day, place your plants in a protected area away from wind and sun for a few hours. Each day, lengthen their time outside. After a few days, you can leave the plants outside overnight, unless there is a threat of a frost. It's good to move the plants each day, exposing them to a little more sun and wind each move.

The last transplanting is to the garden. Dig a trench 5 or 6 inches deep. Place some seasoned manure or commercial fertilizer in the trench, and cover it with 2–3 inches of soil.

If you are planning to stake your tomatoes, put the stakes in now. Carefully remove the tomato plant from its pot, and pinch off all but the top leaves. Lay the tomato down in the trench. Cover the remaining stem with 2–3 inches of soil and pat it down firmly. Bring additional soil up around the leaves to prop the leaves in an upright position. Water the plant well. There should be 16–20 inches between the tomato plants.

Once the soil has warmed up, it is a good idea to spread a thick layer of mulch—hay, leaves, straw, even newspapers—around the plants. The best time to mulch is after a heavy rain. The mulch will help the soil retain moisture and will keep weeds down.

For more information on growing tomatoes you may want to consult *Down-to-Earth Vegetable Gardening Know-How* by Dick Raymond (Garden Way Publishing).

Now with your tomato plants in the garden, the summer weather promising a rich harvest, enjoy. And may all your harvests be bountiful—because really there's no such thing as "Too many tomatoes!"

Preserving
the Harvest

It's the middle of the harvest season, and to-matoes are ripening at a furious pace. You couldn't possibly eat all the tomatoes that are ready in your garden, even if you prepare recipes from this book every night this week. What are you going to do?

Tomatoes are among the easiest vegetables to put by. You have the option of canning tomatoes, which I think gives you the best end product, or you can freeze them. I always dry some of my tomatoes to use in soup and gravy bases, and even in dip mixes. In chapter 2, I will share my recipes for preserving red and green tomatoes in pickles, relishes, jellies, and marmalades.

If you have never preserved any of your garden surplus before, start with tomatoes. Rows of home-processed canning jars filled with tomatoes will come in handy all year and will help you remember and enjoy the fruits of your labor long after the garden has been put to bed.

Since canning is the best way to preserve tomatoes, let's start there.

Canning

Canning seems more difficult than it really is. With careful planning, and scrupulous attention paid to cleanliness, you will have success with your canning, even if you are a beginner. I will give step-by-step instructions to guide you through canning tomatoes. For more detailed information on preserving tomatoes and other fruits and vegetables, you may want to consult *Keeping the Harvest* by Nancy Chioffi and Gretchen Mead or *The Busy Person's Guide to Preserving Food* by Janet Chadwick.

Equipment

For equipment you will need a supply of quart and pint canning jars with the 2-piece screw bands and lids. To make sauces, juices, and purees, you will need a food mill or strainer. You will also need a boiling water bath or steam canner for processing whole and pureed tomato sauces, catsups, and pickled tomatoes. If you plan to process tomato sauces which have added low-acid vegetables or

meats, you will need a pressure canner. That's the basic equipment. In addition, you will find that you need a blanching kettle, assorted bowls, knives, measuring cups and spoons, a nonmetallic spatula for expelling bubbles, clean towels, tongs, potholders, and a jar lifter.

Before you begin canning, be sure you have all your equipment ready. Your work area, jars, and utensils should be spotlessly clean.

Canning Whole Tomatoes In a Boiling Water Bath Or Steam Canner

1. Prepare your jars and lids. Wash the jars, check each jar for nicks and cracks. Place the lids in hot water according to the manufacturer's directions.

Preheat water in the blanching kettle for scalding tomatoes and extra water in a tea kettle. Fill the boiling water bath half full with hot tap water and begin heating. Or fill the well of your steam canner with hot water and begin heating.

2. Wash the tomatoes carefully. Remove any that are overripe. These will be too low in acid to safely can in the boiling water bath or steam canner.

3. Place the washed tomatoes in the steam section of your blanching kettle, and dip the steamer into boiling water for ½ minute.

4. Remove the tomatoes to cold water to cool for 1 minute.

5. Peel off the skins. They should come off very easily. Cut off the stem end and any blemishes. If your tomatoes are large, cut into quarters.

6. Pack the tomatoes tightly into clean jars, pushing the tomatoes down so that they are covered by their own juice. Leave ½ inch head space between the top of the juice-covered tomatoes and the top of the jar. Be sure to pack the tomatoes tightly. If you don't fill the jar with enough tomatoes, the tomatoes you pack will float to the top during processing.

7. Add salt if desired: ½ teaspoon per pint, 1 teaspoon per quart.

8. Run a nonmetallic spatula between the inside of each jar and the tomatoes to release any trapped air bubbles.

9. Wipe the jar rims with a clean, damp cloth.

10. Place the lids in position on each jar and tighten the screw bands.

Once blanched for ½ minute and cooled in ice water, tomato skins peel off easily.

Be sure to pack the tomatoes tightly so they do not float.

11. Place the jars on the rack in the preheated boiling water bath canner. The water should not be boiling; if it is, add cold water to cushion the shock of the temperature difference between the cool jars and the hot water. Then lower the jars into the simmering water with a jar lifter. Be careful not to allow the jars to bump against each other. Add boiling water from the tea kettle to cover the jars by 2 inches. Cover the canner and bring the water to a boil.

If you are processing in a steam canner, set the filled jars on the rack over the preheated water. Be sure the jars are not touching. Set the cover in place and watch for steam to flow out of the vent holes in a steady stream for 5 minutes.

12. Process 35 minutes for pints, 45 minutes for quarts.

In a boiling water bath, begin counting time when the water in the canner has started a rolling boil.

In a steam canner, begin counting time when a steady stream of steam has flowed out of the vent holes for a full 5 minutes.

13. When the processing time is completed, remove the jars from the canner. Place them on heavy towels away from drafts. Allow the jars to cool for 12–24 hours.

14. When the jars are cool, remove the screw bands and test the seals. A jar is sealed when the lid is depressed in the center. Wipe the jars clean and store sealed jars in a cool, dark place.

Canning Tomato Juice or Tomato Puree in a Boiling Water Bath or Steam Canner

1. Prepare your jars and lids. Wash the jars, check each jar for nicks and cracks. Place the jars in hot water. Place the lids in hot water according to the manufacturer's directions.

Fill the boiling water bath canner half full with hot tap water and begin heating. Or fill the well of your steam canner with hot tap water and begin heating.

2. Wash the tomatoes carefully. Remove any that are overripe. They will be too low in acid to safely can in boiling water bath or steam canner. Cut off the stem end of the tomatoes, remove any blemishes. Cut into quarters.

3. If you have a food mill or strainer, such as the Squeezo strainer, which can handle raw tomatoes, process the tomatoes through the strainer. Otherwise, cook the tomatoes until slightly soft. Then process through a strainer or food mill.

4. Pour the strained tomatoes into a kettle. Heat to boiling.

5. To can tomato sauce, continue cooking until the sauce is the desired thickness. Otherwise, ladle the hot juice into clean, hot jars, leaving ½ inch head space.

6. Add salt if desired: ½ teaspoon per pint, 1 teaspoon per quart.

7. Wipe the rim of each jar with a clean, damp cloth.

8. Place the lids in position, and tighten the screw bands.

9. Place the jars in the preheated boiling water bath canner. Add boiling water so that the jars are covered by 2 inches of water. Cover the canner and bring the water to a boil.

If you are processing in a steam canner, set the filled jars on the rack in the preheated canner. Be sure the jars are not touching. Set the cover in place and watch for steam to flow out of the vent holes in a steady stream for 5 minutes.

10. Process pints and quarts for 35 minutes.

Begin counting time in a boiling water bath when the water reaches a rolling boil.

Begin counting time in a steam canner when steam has flowed out of the vent holes in a steady stream for 5 minutes.

11. When processing time is up, remove the jars from the canner. Place the jars on heavy towels away from drafts. Allow the jars to cool for 12–24 hours.

12. When the jars are cool, remove the screw bands and test the seals. A jar is sealed when the lid is depressed in the center. Wipe the sealed jars clean. Label each jar and store in a cool, dark place.

Canning Tomato Paste

Tomato purees can be cooked down over low heat to a very thick paste. Prepare the tomatoes as for tomato juice. Cook the tomatoes for 1 hour before straining. Then strain through a food mill or sieve. Add ½ teaspoon of citric acid (or 1 tablespoon of distilled white vinegar) for every 4 cups of puree. Continue cooking, stirring frequently, until the paste reaches the desired consistency. Ladle the hot paste into hot ½-pint canning jars, leaving ½ inch head space. Process in a boiling water bath or steam canner for 35 minutes.

The Squeezo strainer makes fast work of converting raw tomatoes into tomato puree.

A large canning funnel helps you pour hot puree into canning jars without mess.

Pressure Canning Tomato Sauces

Tomatoes sauces that have low-acid vegetables, such as peppers and onions, added to them must be processed in a pressure canner, unless you are making catsup or chili sauce that has extra vinegar added to increase the acidity of the sauce. When pressure canning a tomato sauce with added vegetables, process as long as the lowest-acid vegetable requires. You can freeze sauces if you don't care to pressure can, but herbs tend to fade in flavor or turn bitter during the freezing process.

Freezing Tomatoes

One year some friends of mine were inundated with cherry tomatoes. As the harvest kept coming and coming, tempers got shorter and shorter, and canning stopped being a rewarding task for them. That's when someone got the idea of simply washing the little tomatoes, tray freezing them on cookie sheets, then bagging the frozen tomatoes in plastic bags.

All year long, whenever they cooked tomato soups, sauces, or gravies, they added these frozen cherry tomatoes (which looked and felt like marbles) to the pot. Depending on what was cooking, I often thought the soups and gravies would have been improved if tomatoes had been defrosted and processed through a strainer to eliminate the skins before being added to the pot. But I certainly couldn't improve on their no-fuss way of preserving.

If you have the freezer space, and don't mind sacrificing the convenience of canned tomatoes, freezing is a good way to preserve tomatoes. You can use frozen tomatoes in any recipe that calls for canned tomatoes. Be sure to defrost the tomatoes first. Once the tomatoes have defrosted, it is easy to remove the skins, if desired.

Freezing Whole Tomatoes

1. Begin preheating water in blanching kettle.

2. Wash the tomatoes carefully.

3. Place the washed tomatoes in the steamer section of your blanching kettle, and dip the filled steamer into the boiling water for ½ minute.

4. Remove the tomatoes to cold water for 1 minute, peel, and quarter.

5. Pack the cooled tomatoes into freezer containers, leaving 1 inch head space.

6. Seal and freeze.

Freezing Tomato Puree/Sauces

1. Wash fresh tomatoes carefully. Remove blemishes. Cut into quarters.

2. If you have a food mill or strainer that can handle raw tomatoes, process the tomatoes through the strainer. Otherwise, cook the tomatoes until slightly soft. Then process.

3. Pack the cooled puree in straight-sided freezer containers, leaving a 1-inch head space.

4. Freeze.

When using frozen purees, be sure to defrost the puree in a bowl and use the liquid that separates from the pulp. It will contain valuable vitamins and should not be thrown away. If you cook the sauce down to a thick consistency, cool the sauce, then freeze it, you will have less liquid separating from the sauce.

It is best not to add herbs to sauces before freezing because the herbs often lose flavor or become bitter.

Drying Tomatoes

Drying takes very little of your time, although the process will extend over several hours. Most of that time you are free to be elsewhere, while the dehydrater works for you. I use my dried tomatoes in soups and stews. Some of the dried tomatoes go into the blender to be pulverized into a fine powder, which I use in dip mixes and to make my own dried soup mix.

1. Wash the tomatoes carefully. Remove any that are overripe or blemished.

2. The next step is optional: peeling tomatoes. To do so, place the washed tomatoes in the steamer section of the blancher and immerse in boiling water for ½ minute. Remove from the boiling water and pour into cold water for 1 minute. Peel, cut out cores. If you plan to make a tomato powder for dips it is not necessary to peel the tomatoes.

3. Cut the tomatoes in ¼-inch slices.

4. Dry. In an electrically controlled dehydrater, spread the slices on the tray so the pieces do not touch. Dry 8–10 hours at 120 degrees F., until brittle.

In the sun, spread the slices on ungalvanized metal screens so that the pieces do not touch. Dry in the hot sun with good air circulation, until the tops are dry. Then turn, and dry the other sides. The tomatoes will dry in 1–2 days, if the weather is

To dry tomatoes, cut the slices ¼ inch thick and spread on dehydrator trays.

After 8-10 hours of drying at 120 degrees F., the tomatoes will be dry and brittle. Store in an airtight container.

good. Be sure to bring the trays in at night so that the dew will not rehydrate the slices.

In a homemade or oven dryer, spread the slices on trays so that the pieces do not touch. Dry at 120 degrees F., until the pieces are hard and crisp, turning the slices and rotating the trays once or twice.

5. Store dried tomatoes in airtight jars in small amounts. Keep away from heat.

For more information on drying, consult *Garden Way's Guide to Food Drying* by Phyllis Hobson.

Keeping Green Tomatoes

I can guarantee that after trying some of my recipes for green tomatoes, you will probably consider harvesting green tomatoes, way before the first frost. But when that first frost does hit, if you have many green tomatoes still on the vine, you will be faced with a big storage problem. Here's what I do with my green tomatoes.

Avoid Frost Damage

Those first frosts that hit us here in northern Vermont come very early, but they are usually not too severe. We cover our tomatoes with heavy cloths when we get a frost warning, and usually that saves the tomatoes.

Sorting Green Tomatoes

Inevitably, the killing frost will come. I harvest the green tomatoes and sort by degree of ripeness. Then I place the tomatoes in cardboard boxes, cover with newspapers, and store in the cellar. The warmer the tomatoes are, the sooner they will ripen. At 65–70 degrees F., mature green tomatoes will ripen in about 2 weeks. At 55 degrees, they will ripen in about 4 weeks. Be sure you check the tomatoes frequently, and remove ripe ones and any that are beginning to spoil.

Some people pull the tomato plants up by the

roots and hang them in the garage. The tomatoes gradually ripen on the vine. They also shed leaves, make a mess, and sometimes fall off the vine. I prefer to box my tomatoes.

Freezing Green Tomato Puree

I hate throwing out good food, and I am always afraid that some of my green tomatoes will spoil before they ripen or before I have time to make enough meals, desserts, or green tomato pickles and relishes. And, as I mentioned, I have found some delicious ways to cook with green tomatoes, so I always put by some frozen green tomato puree to use in soups, sauces, and desserts.

1. Wash 10 pounds of green tomatoes. Dice.

2. Combine the tomatoes in a saucepan with 2 cups water and ¼ cup honey. Simmer for 15 minutes, or until the tomatoes are very soft.

3. Blend until smooth. Cool.

4. Freeze in 1-cup batches.

```
1 pound tomatoes = 2 large tomatoes
                 = 3 medium tomatoes
                 = 4 small tomatoes
2 cups chopped tomatoes = 1½ pounds
1 quart chopped tomatoes = 5 large tomatoes
                         = 7 medium tomatoes
                         = 8–10 small tomatoes
25 pounds (1 peck) tomatoes = 10 quarts canned tomatoes
50 pounds (1 bushel) tomatoes = 20 quarts canned tomatoes
```

Red and Green Preserves

Pickles/Relishes and Sauces
Marmalades and Jellies

Dilled Green Cherries

Yield: 6 quarts
Time: 45 minutes

6 quarts green cherry tomatoes
2 cups white vinegar
¾ cup pickling salt
4 tablespoons pickling spice
   in a muslin bag
2 quarts water
12 grape leaves
6 cloves garlic
24 peppercorns
6 dill heads

Wash and de-stem all the tomatoes. Heat the vinegar, salt, pickling spices, and water to boiling. Simmer for 15 minutes.

In the bottom of each clean quart jar, place 2 grape leaves, 1 clove garlic, 4 peppercorns, and 1 dill head. Pack the jars with green tomatoes. Ladle hot brine over the green tomatoes, leaving ½-inch head space. Seal each jar and process in a boiling water bath or steam canner for 10 minutes. Complete seals if necessary. Allow the jars to sit for 6 weeks before serving.

Sweet Green Wheels

Yield: 10 pints
Time: 1½ hours, plus overnight soaking

35–40 small green tomatoes,
    sliced ¼ inch thick
10 small onions, sliced
2 quarts water
½ cup pickling salt
3 cups water
1 cup cider vinegar
2 cinnamon sticks
1 teaspoon whole cloves
2 tablespoons pickling spice
1 quart cider vinegar
3 cups honey

Layer the tomatoes and onions in a large stainless steel or porcelain bowl. Make a brine with 2 quarts water and the salt. Pour the brine over the tomatoes and onions, and soak overnight.

In the morning, drain the tomatoes and onions and rinse well in cold water.

Pour the tomatoes and onions into a large stainless steel pot, and add 3 cups water and 1 cup of vinegar. Simmer the tomatoes for about 1 hour, or until they look light in color.

Combine the cloves, cinnamon, and pickling spices in a muslin spice bag. In a separate pot, combine the quart of cider vinegar, honey, and spice bag. Boil for 10 minutes.

Fill the hot sterilized pint jars with the tomato and onion mixture. Pour the hot syrup over the tomatoes, leaving a ½-inch head space. Seal the jars and process for 10 minutes in a boiling water bath or steam canner. Complete the seals, if necessary.

Mustard Jumble

Yield: Approximately 7 quarts

Time: 2½ hours, plus overnight soaking

3 quarts green tomatoes,
   coarsely chopped

4 cucumbers, diced with peel

4 small zucchinis, diced with peel

1 head cauliflower, cut in small florets

1 bunch celery, diced on the diagonal

6 red peppers, coarsely chopped

1 quart carrots, cut in ¼-inch circles

1 quart tiny white onions, peeled

1½ cups pickling salt

⅔ cup cider vinegar

4 cups white sugar

1 cup unbleached flour

4 tablespoons Dijon-style mustard

2 tablespoons turmeric

¼ teaspoon cayenne pepper

1 cup white vinegar

This is the pickle to make at the height of the season when you are inundated with garden surplus. It is the most colorful pickle I make.

Mix all the vegetables in a large stainless steel bowl. Sprinkle the salt over the vegetables, mix, and cover the bowl loosely with a towel. Let the vegetables stand overnight.

In the morning, drain the vegetables. Then pour them into a large saucepan and cover with water. Boil for 15 minutes. Drain again.

In the meantime, heat the cider vinegar and sugar in a large stainless steel pot. Mix the flour, mustard, spices, and 1 cup of white vinegar together to form a smooth paste. Add the paste to the vinegar and sugar. Then add the vegetables and heat to boiling. Ladle the vegetables into sterilized hot pint or quart jars, leaving a ½-inch head space, and seal. Process in a boiling water bath or steam canner: pints for 5 minutes and quarts for 10 minutes. Complete seals if necessary.

Virginia Relish

Yield: 10 pints
Time: 2 hours, plus overnight soaking

3 quarts green tomatoes,
  coarsely chopped
3 tablespoons salt
8 cups white cabbage, finely chopped
2 cups sweet red peppers, chopped
1 large onion, finely chopped
4½ cups cider vinegar
2½ cups honey
2 teaspoons celery seed
8 cardamom seeds
2 teaspoons allspice
2 teaspoons mustard seed
2 teaspoons cinnamon

Place the chopped tomatoes in a stainless steel bowl and sprinkle with the salt. Cover loosely with a towel and allow the tomatoes to stand overnight. In the morning, drain the tomatoes well and combine them in a large stainless steel pot with the cabbage, peppers, onion, vinegar, and honey. Simmer for 30 minutes.

Tie the celery seed, cardamom seeds, allspice, and mustard seed in a muslin bag. Put the spice bag in the pot. Add the cinnamon and continue to simmer for 1 hour, or until the mixture thickens. Ladle the relish into sterilized, hot pint jars, leaving a ½-inch head space. Seal and process for 5 minutes in a boiling water bath or steam canner. Complete seals if necessary.

Pennsylvania Chow-Chow

Yield: 10½ pints
Time: 2 hours, plus overnight soaking

½ cup pickling salt
5 quarts green tomatoes, finely chopped
3 cups sweet red peppers, finely chopped
3 quarts onions, finely chopped
5 cups cider vinegar
2 cups white sugar
2 cups honey
½ cup cornstarch
1 teaspoon dried mustard
1 teaspoon turmeric
1 teaspoon curry powder
⅓ cup cider vinegar

Place the vegetables in a stainless steel bowl. Sprinkle the salt over the chopped vegetables and mix together. Cover loosely with a towel, and allow to sit overnight. The next day, drain the vegetables, and place them in a large stainless steel pot with 5 cups of vinegar and the sugar and honey. Cook for 1 hour over medium heat, stirring occasionally.

Make a paste with the cornstarch, mustard, turmeric, curry powder, and ⅓ cup vinegar. Pour this paste into the vegetables, and cook on low heat for 15 minutes. The chow-chow will become quite thick. Ladle the mixture into sterilized, hot pint jars, leaving a ½-inch head space, seal, and process for 5 minutes in a boiling water bath or steam canner. Complete seals if necessary.

Three medium-size onions equal 3 cups sliced or chopped onion, or about 1 pound.

Chinese Tomato Relish

Yield: 4 quarts
Time: 45 minutes

3 cups cider vinegar
2 cups brown sugar, packed
2 tablespoons pickling salt
1 tablespoon ground ginger
1 teaspoon nutmeg
2 teaspoons coriander
4 tablespoons pickling spice,
   tied in a muslin bag
5 quarts barely ripe (pink-green) tomatoes,
   cut in 1-inch chunks
1 bunch celery, diced
1 large onion, coarsely chopped
4 cloves garlic, minced
⅓ cup cornstarch
1 cup cold water

This is a terrific condiment for a Chinese dinner. Or try it heated as a sauce for rice or pork.

In a large stainless steel pot, make a syrup of the vinegar, brown sugar, salt, ginger, nutmeg, and coriander. Add the spice bag. Boil for 5 minutes. Add the tomatoes, celery, onion, and garlic.

In a separate bowl, mix the cornstarch and water. When all the lumps are dissolved, add the cornstarch mixture to the tomatoes and continue boiling for 10 minutes.

Pack the hot relish into 4 hot quart jars or 8 pint jars, leaving a ½-inch head space. Seal and process in a boiling water bath or steam canner: 10 minutes for quarts, 5 minutes for pints.

Old-Time Hot Dog Relish

Yield: 12 ½-pint jars
Time: 3⅓ hours

10 large red tomatoes, peeled
9 peaches, peeled
2 sweet red peppers
3 large onions
1½ cups cider vinegar
1½ cups honey
1 tablespoon pickling salt
4 tablespoons whole pickling spices

Finely chop the tomatoes, peaches, red peppers, and onions. Combine the fruits and vegetables in a large pot; add the vinegar, honey, and salt. Tie the pickling spices in a muslin bag and add it to the pot. Cook the mixture uncovered on medium heat for at least 2 hours, stirring occasionally, until it thickens. Remove the spice bag.

Ladle the hot relish into sterilized hot ½-pint jars, leaving a ½-inch head space, and seal. Process for 10 minutes in a boiling water bath or steam canner. Complete seals if necessary.

Fruited Red Tomato Chutney

Yield: 11 ½-pint jars
Time: 2 hours

12 medium-size red tomatoes, peeled
5 large green sour apples
1 large onion, diced
2 cloves garlic, minced
1½ cups raisins
1 cup dried apricots, diced
1 cup cider vinegar
2 teaspoons salt
1 teaspoon cinnamon
dash cayenne pepper
⅓ cup candied ginger, finely diced

In a large, heavy-bottomed saucepan, heat all the ingredients, stirring frequently. Simmer uncovered for about 1½ hours. The chutney will get quite thick.

When the chutney has thickened, keep it simmering. Ladle hot chutney into sterilized, hot ½-pint canning jars, leaving a ½-inch head space, and seal. Process jars in a boiling water bath or steam canner for 10 minutes.

To peel a tomato, grasp it with tongs or with a long carving fork and dip it into boiling water for several seconds. Then plunge it into cold water. The skin will be loosened, and can be peeled off easily.

Red Tomato Catsup

Yield: 5 ½-pint jars
Time: 2 hours

6 pounds red tomatoes
1 large onion, chopped
1 sweet red pepper, diced
2 teaspoons celery seed
1½ teaspoons allspice berries
1 teaspoon mustard seed
2 cinnamon sticks
½ cup honey
1 tablespoon salt
1¼ cup cider vinegar

Italian plum tomatoes make the best catsup because they have less water than other varieties.

In a heavy saucepan, combine the tomatoes, onion, and pepper, and simmer uncovered for 30 minutes on medium heat. Puree this mixture in a blender and return to the pot. Tie the celery seed, allspice berries, mustard seed, and cinnamon sticks in a muslin bag and add it to the pot. Add the honey, salt, and the vinegar. Simmer the pot for 20–30 minutes more, stirring occasionally. The catsup will thicken. Remove the spice bag.

Ladle the hot catsup into sterilized, hot ½-pint canning jars, leaving a ½-inch head space. Seal. Process in a boiling water bath for 10 minutes. Complete seals if necessary.

Green Gringo Taco Sauce

Yield: 6 ½-pint jars

Time: 3 hours

2 quarts green tomatoes, finely chopped (save the juice)

1½ cups cider vinegar

2 medium-size onions, chopped

1½ teaspoons pickling salt

½ teaspoon black pepper

½ teaspoon cayenne pepper

2 tablespoons dry mustard

This is a mild taco sauce for sensitive palates.

Combine all the ingredients in a heavy-bottomed saucepan, and simmer for 2 hours. Pour the mixture into a blender and process until it is smooth. Taste and correct the seasoning.

Fill sterilized, hot ½-pint jars, leaving a ½-inch head space. Seal and process in a boiling water bath or steam canner for 10 minutes. Complete seals if necessary. Instead of canning this relish, you can freeze it, if you prefer.

Green Tomato Mincemeat

Yield: 10 pints
Time: 3 hours

8 cups green tomatoes, finely chopped
1-1½ cups cider (approximately)
8 cups unpeeled apples, finely chopped
1½ pounds raisins
1 pound dates, chopped
2 cups honey
1⅓ cups cider vinegar
2 tablespoons cinnamon
1 teaspoon allspice
2 teaspoons cloves
¼ teaspoon pepper
2 tablespoons orange peel
⅔ cup vegetable oil

Mincemeat is a favorite old-time preserve that is most often seen at Thanksgiving in mincemeat pies, but it is a shame not to use this delicious, versatile green tomato preserve in as many ways as possible. In Chapter 3, I've included recipes for mincemeat cookies, cakes, and muffins; but if it is pies you are planning to make, figure that 1 pint will fill an 8-inch pie.

Drain the chopped tomatoes for 5–10 minutes, and replace the juice with the same amount of apple cider. Mix all the ingredients, except the oil, in a large stainless steel pot. Simmer until thick—approximately 2 hours. Add the oil and stir well.

Ladle into sterilized, hot pint canning jars, leaving a ½-inch head space. Seal and process for 25 minutes in a boiling water bath or steam canner. Complete seals if necessary. You can freeze mincemeat instead of canning, if you wish.

Gladys's Old-Fashioned Green Tomato Mincemeat with Suet

Yield: 8–10 pints

Time: 2½ hours

8–10 pounds green tomatoes,
 chopped and drained

3 pounds apples, chopped

2 pounds raisins, washed and drained

3 cups sugar

1 cup cider vinegar

1 cup strong coffee

1 cup suet, chopped

1 cup apple jelly or cider

1 teaspoon salt

2 teaspoons cinnamon

1 teaspoon nutmeg

¾ teaspoon ground cloves

Combine all the ingredients in a large stainless steel pot and simmer until thick, about 1½ hours, stirring often.

Ladle the mincement into sterilized, hot pint canning jars, leaving a ½-inch head space. Seal. Process in a boiling water bath or steam canner for 25 minutes. Complete seals if necessary. You can freeze mincemeat, if you don't wish to can.

Teatime Marmalade

Yield: 8 ½-pint jars

Time: 2 hours

3 medium-size lemons

½ cup water

5 pounds green tomatoes, quartered and thinly sliced

4 cups sugar

1 cup light brown sugar

1 tablespoon canning salt

2 cinnamon sticks

Peel the skin off the lemon in thin strips, so that very little white membrane is included, and sliver the skin as thinly as possible.

In a small saucepan, heat the water to boiling and add the lemon peel. Simmer the peel for 10 minutes and drain.

Trim off all the white membrane on the lemons, cut the lemons into thin slices. Quarter the slices, and remove the seeds. Combine the cooked peel, lemons, tomatoes, sugars, salt, and cinnamon in a large pot. Cook the marmalade uncovered over medium heat for 50 minutes, or until very thick. Stir the marmalade occasionally. Remove the cinnamon sticks.

Ladle into sterilized, hot ½-pint jars, leaving a ½-inch head space. Seal and process in a boiling water bath or steam canner for 10 minutes. Complete seals if necessary.

Tomato Herb Jelly

Yield: 4 ½-pint jars
Time: 45 minutes

¼ cup chopped fresh sage or
  ¼ cup crushed fresh marjoram
¼ cup crushed fresh thyme
1 cup water
1 cup red tomato juice
dash Tabasco sauce
½ cup lemon juice, strained
2 cups honey
2 cups sugar
1 bottle liquid pectin

This jelly is delicious served with pork and lamb.

Combine the herbs in a small saucepan with the water. Simmer gently for 5 minutes and set aside to steep for 30 minutes.

Combine the tomato juice, Tabasco, lemon juice, honey, and sugar in a large stainless steel pot.

Strain the herbs and save. Add the herb water to the tomato mixture. Return 2 teaspoons of the herbs to the pot and discard the rest.

Bring the jelly to a hard boil. Add the pectin and boil hard for 1 minute. Remove the pot from the heat and skim off any froth. Ladle the jelly into sterilized, hot ½-pint jars, leaving a ½-inch head space, and seal.

Lemon Tomato Jelly

Yield: 4 ½-pint jars
Time: 30 minutes

2 cups red tomato juice
½ cup lemon juice, strained
2 cups sugar
2 cinnamon sticks
2 cups honey
½ bottle liquid pectin

Jars of this red-colored jelly look very festive when topped with decorative squares of green fabric and tied with ribbons—perfect Christmas gifts.

In a large stainless steel pot, combine the tomato juice, lemon juice, sugar, cinnamon sticks, and honey. Bring to a rolling boil. Remove the jelly from the heat and let it sit for 15 minutes. Remove the cinnamon sticks and discard. Add the pectin and return the jelly to a hard boil for 1 minute. Remove the pot from the heat and skim off any froth. Ladle the jelly into sterilized, hot ½-pint jars, leaving a ½-inch head space, and seal.

Green Tomato Delights

Appetizers/Soups
Dressings and Salads
Side Dishes/Main Dishes
Breads and Desserts

Bacon Cheese Ball

Yield: 8–10 servings
Time: 20 minutes

1 pound cream cheese
½ pound bacon, browned and crumbled
4 scallions, finely diced
1 green tomato, finely chopped
1 cup finely chopped walnuts

Mix all the ingredients, except the nuts, and form into a ball. Cover with walnuts. Place the ball in the refrigerator for at least 1 hour before serving.

Green Tomato Nacho Bean Dip

Yield: 8 servings
Time: ½ hour

2 tablespoons vegetable oil
1 medium-size onion, diced
2 green tomatoes, diced
2 cups cooked red kidney beans
1½ cups Red Taco Sauce (p. 91)
salt and pepper to taste
1½ cups shredded cheddar cheese

In a large frying pan, heat 2 tablespoons oil, and sauté the onion until it appears translucent. Add the green tomatoes and cook for 5 minutes. Add the beans, 1 cup of Taco Sauce, and salt and pepper. Continue cooking for 15 minutes. Mash some of the beans with the back of a spoon to make the mixture look pasty. Transfer the mixture to a shallow baking dish. Sprinkle the cheese and the remaining ½ cup Taco Sauce on top. Broil for 5 minutes (or until the cheese is bubbly). Serve with corn chips.

Spiced Carrot Soup

Yield: 6 servings
Time: 45 minutes

1 pound carrots, diced
4 large green tomatoes, diced
1 medium-size onion, diced
7 cups chicken or vegetable broth
3 teaspoons ginger
1 teaspoon nutmeg
1 cup heavy cream
1 tablespoon honey
salt and pepper to taste

In a medium-size soup pot, combine the vegetables and the broth. Simmer until the vegetables are tender, about 15 minutes. Cool slightly and put the mixture through a blender until smooth. Return the mixture to the soup pot and add the remaining ingredients. Reheat and serve.

1 tablespoon shredded fresh ginger root can be substituted for ½ teaspoon ground ginger.

Sweet Potato Soup

Yield: 8 servings

Time: 1½ hours

5 sweet potatoes

5 green tomatoes

¼ cup water

4 cups milk

2 tablespoons vegetable oil

1 large onion, diced

2 cups zucchini, diced

3 cloves garlic, minced

1½ cups green peas (fresh or frozen)

1 tablespoon honey

1 pound cheddar cheese, shredded

3 tablespoons tamari (soy sauce) or salt to taste

Cut the sweet potatoes into pieces, but do not peel. Place the potatoes in a pot with enough water to cover and boil until they are soft. Drain and skin them.

In the meantime, dice the green tomatoes and put them in a soup pot with ¼ cup water. Cook the tomatoes until soft.

Puree the sweet potatoes and green tomatoes together in a blender, adding as much milk as necessary to achieve a soup consistency. Return the puree to the soup pot.

In another pan, heat the oil, and sauté the onion and the zucchini until they are tender crisp. Add to the soup pot. Add the remaining ingredients, and correct the seasoning. Reheat the soup on low.

> Vegetables that are steamed or sautéed just long enough to change their color, but not until they are soft, are called "tender crisp."

Russian Green Tomato Borscht

Yield: 6 servings
Time: 2 hours

2 tablespoons bacon fat or vegetable oil
5 green tomatoes, finely chopped
1 medium-size onion, finely diced
3 cups cooked beets, diced
1 pound beef, cooked and diced
6 cups beef broth
4 tablespoons honey
3 tablespoons lemon juice
salt and pepper to taste
½ pound bacon, cooked and diced
  (optional)
½ cup sour cream

In a medium-size soup pot, heat the bacon fat, and sauté the green tomatoes, onion, beets, and beef for 10 minutes. Add the remaining ingredients, except sour cream, and simmer for 1½ hours. Taste and adjust seasoning. Garnish with sour cream.

Slavic Tomato Mushroom Soup

Yield: 8 servings
Time: 1¼ hours

2 tablespoons vegetable oil
2 large onions, diced
1½ pounds fresh mushrooms, sliced
2 potatoes, diced
2 cups finely diced green tomatoes
6 cups water
2 teaspoons dill
2 tablespoons Hungarian paprika
¼ cup tamari (soy sauce)
3 tablespoons butter
3 tablespoons unbleached flour
2 cups milk
salt and pepper to taste
¼ cup lemon juice
1 cup sour cream

In a large soup pot, heat the vegetable oil, and sauté the onions and mushrooms for 5 minutes. Add the potatoes and tomatoes and cook for 5 minutes. Add the water, dill, paprika, and tamari. Cover and cook for 20 minutes.

In the meantime, melt the butter in a small saucepan, and add the flour, stirring constantly to smooth out any lumps. Slowly add the milk, and continue stirring to make a thick white sauce. Add some of the broth from the soup pot to the white sauce and pour this mixture back into the soup pot. Cover the pot and simmer for 10 minutes. Just before serving, add salt, pepper, lemon juice, and sour cream.

Green Goddess Salad Dressing

Yield: 1¾ cups
Time: 10 minutes

1 cup green tomatoes, finely chopped
1 tablespoon chopped fresh tarragon or
  2 teaspoons dried tarragon
¼ cup finely diced chives
½ cup finely chopped fresh parsley
2 cloves garlic, minced
2 tablespoons tarragon vinegar
1½ cups mayonnaise
salt and pepper to taste

This recipe is especially easy to make with a food processor. The tomatoes and herbs can be chopped together.

Combine the tomatoes, herbs, and garlic. Mix well. Add the vinegar, mayonnaise, and salt and pepper, and mix lightly.

Sweet & Sour Salad Dressing

Yield: 3 cups
Time: 10 minutes

2 cups salad oil
½ cup wine vinegar
½ teaspoon salt
pepper to taste
3 small green tomatoes, chopped
1 heaping tablespoon Dijon-style mustard
2 tablespoons honey, or to taste

Combine all ingredients in a blender and blend until smooth.

Green Mountain Coleslaw

Yield: 8–10 servings
**Time: 25 minutes (plus 1 hour
     standing time)**

1 large head green cabbage, shredded
3 large carrots, grated
**3 medium-size green tomatoes, finely
     diced**
3 medium-size apples, diced
½ cup chopped walnuts
¾ cup raisins
1½ cups mayonnaise
¼ cup maple syrup
¼ cup lemon juice
¾ teaspoon nutmeg

In a large salad bowl, combine the cabbage, carrots, tomatoes, apples, walnuts, and raisins.

In a separate bowl, whisk together the remaining ingredients and pour over the salad. Mix well and let the salad sit for an hour before serving.

Carrot Confetti in Pineapple Boats

Yield: 6–8 servings
Time: 30 minutes

3 cups grated carrots
**2 medium-size green tomatoes, finely
  chopped**
½ cup raisins
½ cup chopped walnuts
1 fresh pineapple
⅓ cup honey or maple syrup
½ cup lemon juice
¼ cup chopped parsley

Mix together the carrots, tomatoes, raisins, and walnuts.

Cut the pineapple in half horizontally, leaving the green top attached. Cut out the tough core of the pineapple, and scoop out the pulp, leaving "boats" of the pineapple shells. Chop the pulp and add to the carrot mixture.

In a separate bowl, mix the honey and lemon juice. Combine the liquid with the carrot mixture. Pile the carrot salad into the empty pineapple shells. Decorate with the chopped parsley.

> You can tell that a pineapple is ripe by its smell. Also, the leaves will pull out easily. Beware of pineapples that are completely yellow; they are usually overripe.

Fried Green Tomatoes

Time: 20 minutes

1 medium-size green tomato per person
eggs and milk
2 tablespoons vegetable oil

Breadings
(approximately ½ cup per tomato):
Cornmeal, bread crumbs, wheat germ, or
flour

Seasonings (½ teaspoon per tomato):
Parmesan cheese, salt and pepper, Italian
herbs, caraway seeds, curry powder,
sesame seeds, or chili powder

Oils (2 tablespoons per tomato):
Bacon grease, vegetable oil, butter, or
olive oil

Fried green tomatoes are one of the few dishes that are "standard" for green tomatoes. You can make your green tomatoes very "unstandard" by varying the breading or seasonings. Another variation that will be appreciated is serving the Fried Green Tomatoes with a white sauce seasoned with curry powder or Parmesan cheese.

Beat the eggs and combine with the milk (approximately 2 tablespoons milk per egg). Combine the breading with the seasoning of your choice.

Slice the tomatoes into ½-inch slices. Dip the tomatoes into the egg and milk mixture. Cover with the seasoned breading. Heat the oil and fry the tomatoes until they are golden crispy brown on both sides. Serve immediately.

Log Cabin Potatoes

Yield: 8 servings
Time: 1½ hours

8 medium-sized unpeeled potatoes, thinly
   sliced
3 large green tomatoes, thinly sliced
1 medium-size onion, diced
1 cup unbleached flour
1 pound cheddar cheese, grated
¼ pound bacon, browned and crumbled
salt and pepper to taste
½ cup milk

Preheat the oven to 350 degrees F.

Butter a large baking dish. Put a layer of potatoes on the bottom and cover with a layer of green tomatoes. Sprinkle on a little bit of onion, flour, cheddar cheese, bacon, and salt and pepper. Continue layering until the dish is full, ending with a layer of cheese. Pour the milk over the top and bake for 1 hour, or until potatoes are cooked and bubbling brown on top.

Vermont Sweet Potatoes

Yield: 8 servings
Time: 1 hour

6 sweet potatoes, cooked and thinly sliced
5 green tomatoes, sliced
4 tablespoons butter
½ cup maple syrup
½ teaspoon salt
1 cup coconut, shredded

Preheat the oven to 350 degrees F.

Butter a casserole dish. Cover the bottom of the dish with 1 layer of sweet potatoes and 1 layer of green tomato slices. Continue layering until the casserole is full.

Melt the butter and add the maple syrup and salt. Pour this mixture over the potatoes and tomatoes. Sprinkle the coconut on top. Bake for 30 minutes uncovered or until bubbling hot.

Yellow Confetti Rice

Yield: 6 servings
Time: 50 minutes

4 cups water
½ teaspoon salt
2 cups brown rice
⅛ teaspoon saffron or
　½ teaspoon turmeric
1 small onion, diced
1 medium-size green tomato, diced
½ cup raisins
½ cup chopped walnuts

Serve this dish with lamb or pork or any dish that calls for cooked rice.

Bring the water to boil in a medium-size saucepan. Add the salt, rice, saffron or turmeric, and onion. Cover the pot and reduce the heat. Simmer for 30 minutes.

Add the green tomatoes to the pot, but do not stir. Continue cooking for 15 more minutes. The rice should be tender, and the water absorbed. Remove the pot from the heat and stir in the raisins and nuts, using a fork to fluff up the rice. Serve hot.

Jubilee Carrots

Yield: 8 servings
Time: 10 minutes

1 pound carrots, sliced
½ cup water
2 medium-size green tomatoes, diced
2 tablespoons butter
¼ cup brown sugar
1 tablespoon Dijon-style mustard
salt and pepper to taste

In a medium-size saucepan, simmer the carrots in the water for 5 minutes uncovered. Add the tomatoes and cook for 5 minutes. Add the remaining ingredients and cook for another 3 minutes. Serve hot.

Spaghetti Squash Supreme

Yield: 8 servings
Time: 1½ hours

1 medium-size spaghetti squash
1 medium-size eggplant
3 medium-size green tomatoes
4 tablespoons olive oil
3 cloves garlic, minced
½ cup chopped parsley
salt and pepper to taste
2 tablespoons butter
½ pound mushrooms, sliced
1 cup grated cheddar cheese

Preheat the oven to 400 degrees F.

Pierce the skin of the spaghetti squash in several places with a fork, and bake the squash for 45 minutes, or until it feels soft when you press.

In the meantime, peel the eggplant and cut it into small cubes. Chop the green tomatoes into ½-inch cubes, and drain in a colander for 5–10 minutes.

Heat the olive oil in a large frying pan, and sauté the eggplant for 5 minutes. Add the green tomatoes and the garlic, and cook for 10 minutes. Add the parsley and salt and pepper, and set aside until the squash is done. When the squash is soft, cut it in half lengthwise, and scoop out the seeds. Use a fork to remove all the pulp from the squash. It will come out looking like spaghetti.

In a small frying pan, melt the butter, and sauté the mushrooms until tender. Drain off the liquid, add the squash, and heat. Meanwhile, reheat the tomato-eggplant mixture.

Place the squash-mushroom mixture in a casserole dish. Sprinkle with the cheese. Pour the tomato-eggplant mixture on top. Serve at once.

Pasta Charlotte

Yield: **6 servings**
Time: **1 hour**

4 cups chopped green tomatoes
½ cup white wine
2 tablespoons butter
1 large onion, diced
1 cup heavy cream
¼ cup fresh finely chopped basil
1 teaspoon honey
salt and pepper to taste
2 pounds spiral noodles
Parmesan cheese to garnish
1 cup sliced black olives

In a large saucepan, combine the tomatoes and wine. Cook uncovered for 45 minutes, stirring occasionally. Puree the mixture in a blender and set aside.

In a large saucepan, melt the butter, and sauté the onion. Add the tomato puree, cream, basil, honey, salt and pepper. Keep the sauce warm.

In the meantime, cook the noodles and drain. Arrange the noodles on a serving dish, and pour the sauce over the noodles. Sprinkle with Parmesan cheese and black olives. Serve at once.

Old-Fashioned Chicken Pot Pie

Yield: 6 servings
Time: 1½ hours

Filling:

4 tablespoons butter
4 large green tomatoes, chopped
3 large carrots, diced
1 large onion, diced
4 stalks celery, diced
4 cups diced cooked chicken
½ cup butter
½ cup unbleached flour
3 cups milk
¼ cup fresh chopped basil or
   2 teaspoons dried basil
salt and pepper to taste

Biscuit Dough:

1 cup whipping cream
2 cups unbleached flour
½ teaspoon salt
4 teaspoons baking powder
½ teaspoon cream of tartar
⅓ cup milk

Preheat the oven to 350 degrees F.

In a large frying pan, melt the 4 tablespoons of butter, and sauté the vegetables for 10 minutes. Transfer them to a bowl with the chicken. In the same pan, melt the ½ cup butter, and add ½ cup flour, stirring constantly to get rid of lumps. Slowly stir in the milk to make a thick white sauce. Add the sauce and the basil to the chicken and vegetables. Add salt and pepper to taste. Transfer the chicken mixture to a greased casserole dish.

To make the biscuit dough, whip the cream. Sift together the flour, salt, baking powder, and cream of tartar. Fold the dry ingredients and milk in with the cream, stir just enough to mix them. Drop the biscuit dough by spoonfuls onto the casserole to cover the top. Bake for 25 minutes or until the biscuits begin to brown.

Chicken Dijon with Green Tomatoes

Yield: 6 servings
Time: 1½ hours

2 tablespoons vegetable oil
2 tablespoons butter
6 pieces of chicken
½ cup butter
¾ cup unbleached flour
1½ cups milk
¾ cup white wine
4 tablespoons Dijon-style mustard
2 tablespoons butter
5 green tomatoes, finely diced
salt and pepper to taste

This dish is a good one for a busy person. It can be made early in the day, refrigerated, and reheated for about ¾ hour before serving.

Preheat the oven to 400 degrees F.

In a large frying pan, heat the oil and 1 tablespoon of butter. Brown the chicken pieces on both sides, then transfer them to a baking dish.

Add ½ cup butter to the frying pan and melt. Mix the flour into the butter, smoothing out any lumps. Slowly add the milk, a little at a time. Continue stirring to get rid of any lumps. Add the wine and the mustard. Cook the sauce for 10 minutes on a low heat.

In a separate pan, melt 2 tablespoons of butter, and sauté the tomatoes for 10 minutes. Add the tomatoes to the mustard sauce. Correct the seasoning.

Pour the sauce on top of the chicken and bake for 30–45 minutes. Serve over rice or noodles.

Green Tomato Tarragon Chicken

Yield: 6 servings
Time: 1½ hours

2 tablespoons vegetable oil
2 tablespoons butter
6 pieces of chicken
4 cups diced green tomatoes
½ cup water
1 tablespoon honey
1 cup heavy cream
1 tablespoon tarragon
salt and pepper to taste
3 tablespoons Parmesan cheese

Preheat the oven to 350 degrees F.

In a large frying pan, heat the oil and the butter, and brown the chicken on both sides. Remove the chicken and place in a baking dish.

In a medium-size saucepan, combine the tomatoes, water, and honey, and cook for ½ hour or until the tomatoes are soft. Puree the tomatoes with the cooking liquid in a blender. Add the cream, tarragon, and salt and pepper. Pour this sauce over the chicken and bake for 45–60 minutes. Serve over rice or noodles and garnish with Parmesan cheese.

Curried Chicken

Yield: 6 servings
Time: 45 minutes

3 whole boneless chicken breasts
1½ cups unbleached flour
¼ cup vegetable oil
2 tablespoons vegetable oil
5 large green tomatoes, diced
1 large onion, sliced
½ cup apple cider
2 tablespoons curry powder
½ teaspoon cumin
½ teaspoon coriander
dash cayenne pepper
1 cup heavy cream
salt to taste

Cut the chicken into 1-inch cubes and dredge with flour. Heat ¼ cup oil in a large frying pan. Brown the chicken in the oil for 10 minutes. Remove the chicken from the pan. Add 2 more tablespoons of oil to the pan, and sauté the vegetables for 10 minutes. Mix the cider, curry powder, cumin, coriander, and cayenne, and add to the vegetables. Add the chicken and the cream. Add salt to taste. Simmer for 10 minutes to let the flavors mingle. Serve over rice.

Stroganoff Stew

Yield: 6 servings
Time: 2 hours

½ pound bacon, diced
¾ cup unbleached flour
1 teaspoon salt
½ teaspoon pepper
2 teaspoons mixed Italian herbs
1½ pounds stew beef
5 green tomatoes, finely chopped
1 medium-size onion, chopped
3 cups water
1 pound mushrooms, sliced
2 cups sour cream
3 tablespoons cornstarch mixed with 3
   tablespoons water (optional)
¼ cup tamari (soy sauce) or to taste
¼ cup chopped fresh basil or
   2 tablespoons dried basil

Cook the bacon in a heavy-bottomed pot. Remove the bacon pieces, but leave the bacon grease in the pot.

Mix together the flour, salt, pepper, and herbs. Dredge the stew beef in the flour mixture. Sauté the meat in the bacon grease until browned. Add the tomatoes, onion, water, and any remaining flour. Cover the pot and simmer for 1½ hours. Then add the mushrooms and sour cream. (For a thicker gravy, make a paste by combining the cornstarch with the water and add to the sour cream, before adding the sour cream to the pot.) Heat on low for 15 minutes and adjust the seasoning with the tamari and basil. Serve over buttered noodles.

> You can make your own mix of Italian herbs by combining equal portions of basil, oregano, thyme, marjoram, and rosemary.

Sesame Green Tomato Quiche

Yield: 8 servings
Time: 1¾ hours

Crust:

1 cup unbleached flour
¼ teaspoon salt
¼ cup sesame seeds
⅓ cup vegetable shortening or butter
2–4 tablespoons water

Filling:

2 tablespoons vegetable oil
1 medium-size onion, diced
2 medium-size green tomatoes, finely
   diced
2 cups sliced mushrooms
1½ cups diced cooked ham (optional)
4 eggs
2 cups light cream
dash nutmeg
½ teaspoon salt
dash pepper
½ pound Swiss cheese, grated
2 tablespoons sesame seeds

Combine the flour, salt, ¼ cup sesame seeds. Cut the shortening into the flour and blend with a fork until the mixture resembles small peas. Add the water and mix until the dough forms a ball. Refrigerate the dough for ½ hour.

Meanwhile, heat the oil in a medium-size saucepan, and sauté the onion for 5 minutes. Add the green tomatoes and mushrooms and cook for 3 minutes. Let the vegetables cool slightly.

Preheat the oven to 350 degrees F.

Roll out the pie crust on a lightly floured board. Fit it into a 10-inch pie pan, and flute the edges. Spread the vegetables over the bottom crust. Sprinkle the diced ham over the vegetables.

In a mixing bowl, beat together the eggs, cream, nutmeg, salt, and pepper. Pour this mixture into the pie shell. Sprinkle the grated cheese evenly over the pie and top with sesame seeds. Bake the quiche for 1 hour, or until golden brown on top.

Oatmeal Cookie Bars

Yield: 25 bars
Time: 1 hour

4 cups finely chopped green tomatoes
1 cup dark brown sugar, firmly packed
2 tablespoons lemon juice
1 teaspoon lemon extract
¾ cup butter
1 cup dark brown sugar, firmly packed
1½ cups unbleached flour
½ teaspoon baking soda
½ teaspoon salt
2 cups rolled oats
½ cup chopped walnuts

Preheat the oven to 375 degrees F.

Drain the green tomatoes for 5–10 minutes. Place the tomatoes in a saucepan with 1 cup brown sugar and the lemon juice. Simmer for 15–20 minutes or until the mixture is quite thick. Remove from the heat and stir in the lemon extract. Set aside.

In a mixing bowl, cream the butter. Mix in 1 cup brown sugar, and beat until fluffy. Add the flour, baking soda, and salt to the mixture. Mix well. Stir in the oats and walnuts.

Grease a 9-inch by 12-inch pan and press ⅔ of the oatmeal mixture into the bottom of the pan. Spread the green tomato conserve on next. Sprinkle the remaining oatmeal mixture over the top.

Bake for 30–35 minutes. Cool in the pan and cut into bars.

Variation: 2 cups of Green Tomato Mincemeat (page 28) can be used instead of the green tomato conserve.

Green Tomato Coconut Bars

Yield: 25 bars
Time: 1¼ hours

2 cups unbleached flour, sifted
1 cup butter, at room temperature
¼ cup sugar
1 teaspoon cinnamon
2 cups green tomatoes, finely chopped
3 eggs
1 cup brown sugar, firmly packed
1 cup chopped walnuts
1 lemon peel, grated
2½ cups coconut

Preheat the oven to 350 degrees F.

Cut the butter into the flour with a pastry blender. Add the sugar and cinnamon. Press the mixture into an 8-inch by 12-inch pan and bake for 10 minutes to form a crust.

To make the top layer, mix the remaining ingredients in a bowl. When the crust layer is done, pour on the top layer, and return the pan to the oven to bake for 25 minutes, or until set. Cool slightly and cut into small bars.

Cream Cheese Frosting

Yield: 3½ cups, enough for a 9-inch
double-layer cake
Time: 10 minutes

8 ounces cream cheese
¾ cup butter
4 cups confectioners' sugar

Bring the butter and cream cheese to room temperature and cream together until light and fluffy. Slowly add the confectioners' sugar until well blended, using more sugar if necessary. Frost the top and sides of the cake, sprinkle with nuts if desired.

Spiced Chocolate Cake

Yield: 12 servings
Time: 1½ hours

Cake:

½ cup butter
1 cup sugar
4 eggs, beaten
2 cups Green Tomato Mincemeat (p. 28)
3 cups unbleached flour, sifted
½ cup cocoa
1 teaspoon cinnamon
½ teaspoon nutmeg
2 teaspoons baking soda
1 teaspoon baking powder

Frosting:

1½ cups Cream Cheese Frosting (p. 54)
½ cup cocoa
2 teaspoons cinnamon

Preheat the oven to 350 degrees F.

Cream together the butter and sugar until fluffy. Beat in the eggs and the mincemeat. Sift together the dry ingredients and add to the batter. Beat for 1 minute.

Butter 2 round 9-inch cake pans. Divide the batter between the pans and bake for 40 minutes, or until the cake tests done. To make the frosting, prepare the Cream Cheese Frosting and add to it ½ cup cocoa and 2 teaspoons cinnamon. Frost the cooled cake.

Andrea's Green Tomato Chocolate Cake

Yield: 12–15 servings
Time: 1 hour

⅔ cup butter

1¾ cups sugar

4 ounces unsweetened chocolate, melted

2 eggs

1 teaspoon vanilla extract

½ cup cocoa

2½ cups sifted unbleached flour or
 1½ cups sifted unbleached flour and
 1 cup sifted whole wheat flour

2 teaspoons baking powder

2 teaspoons baking soda

¼ teaspoon salt

1 cup beer

1 cup pureed green tomatoes

¼–½ cup water (optional)

Preheat the oven to 350 degrees F.

Cream together the butter and sugar. Stir in the melted chocolate, then the eggs, one at a time. Add the vanilla.

In another bowl, sift together the cocoa, flour, baking powder, baking soda, and salt.

Add the flour mixture to the butter mixture alternately with the beer and green tomatoes. If the batter appears stiff, add the water. (Sometimes the moisture content of the tomatoes vary.)

Turn the batter into a greased and floured 9-inch by 13-inch baking dish. Bake for 35 minutes.

Ice with your favorite frosting when cooled, or serve plain. This is a very rich, moist cake.

Mini Mincemeat Cakes

Yield: 4 dozen cookies
Time: 1½ hours

½ cup butter
½ cup brown sugar
½ cup honey
1 egg
3 cups unbleached flour, sifted
¼ teaspoon baking soda
1 teaspoon baking powder
1 teaspoon ground cloves
1 teaspoon ground cinnamon
1½ cups Green Tomato Mincemeat
  (p. 28)
1½ cups ground walnuts
½ cup sugar

Preheat the oven to 400 degrees F.

Cream together the butter and brown sugar until fluffy. Beat in the honey and the egg.

In a separate bowl, mix the flour, baking soda, baking powder, and the spices. Add the dry ingredients to the butter mixture, and mix well. Add the mincemeat and stir.

Drop the batter by spoonfuls into a small bowl filled with the walnuts and sugar. Shape the batter into small balls making sure each ball is covered with nuts.

Bake for 10 minutes.

Green Tomato Breads and Desserts/57

Old-Time Bread Pudding With Mincemeat

Yield: 8 servings
Time: 1½ hours

12 slices bread (whole wheat preferred)
6 tablespoons butter
1 cup Green Tomato Mincemeat (p. 28)
4 eggs
1¾ cups milk
½ cup maple syrup or honey

Preheat the oven to 350 degrees F.

Butter the slices of bread on both sides. Line a buttered loaf pan with 8 slices of bread. Spread half of the mincemeat in the pan. Top with 2 slices of bread. Spread the remaining mincemeat in the pan. Cover with 2 more slices of bread.

Beat the eggs, and mix in the milk and maple syrup. Pour the milk mixture over everything. Bake for at least 1 hour or until a knife comes out clean.

When you find yourself with leftover canned mincemeat after preparing a sweet bread, pie, cake, or cookies, freeze the remainder to use another day.

Mincemeat Coffeecake Twist

Yield: 8 servings
Time: 3 hours

⅓ cup milk

2 tablespoons dried baker's yeast

3⅓ cups all-purpose flour or unbleached
flour, sifted

4 tablespoons butter, melted

⅓ cup sugar

½ teaspoon salt

2 eggs

1 cup Green Tomato Mincemeat (p. 28)

Heat the milk until just lukewarm. Mix the yeast and 2 tablespoons of the warm milk until smooth. Then add the remaining milk.

Pour the flour into a large bowl and make a well in the center. Pour the milk and yeast mixture into the well, sprinkle a little flour over the yeast, and cover the bowl with a towel. Set the bowl in a warm place for 15 minutes.

In the meantime, melt the butter, and combine it with the sugar, salt, and eggs. Stir the yeast and flour mixture, and add the butter and sugar mixture. Stir until the flour is all blended in.

On a lightly floured board, knead the dough for about 10 minutes or until it is smooth. Cover with a towel and let the dough rise in a warm place for 30–40 minutes.

Roll out the dough to form a rectangle 18 inches by 12 inches. Spread the mincemeat over the dough. Roll the dough lengthwise, and seal the edges by pressing and pinching them together. Cut the roll in half lengthwise and twist the pieces around each other like a 2-piece braid to make a twisted roll. Place the bread on a greased cookie sheet, and let it rise for 15 minutes. Meanwhile, preheat the oven to 400 degrees F.

Bake for 30 minutes.

Mincemeat Upside-Down Cake

Yield: 12 servings

Time: 1 hour

3 cups Green Tomato Mincemeat (p. 28)
¾ cup butter
1 cup milk
2 eggs, beaten
3 cups unbleached flour, sifted
4 teaspoons baking powder
½ teaspoon salt
1 cup sugar
2 teaspoons cinnamon

Preheat the oven to 400 degrees F.

Melt 4 tablespoons of butter and pour into a bundt pan or ring-shaped cake pan. Coat the pan by tipping it until it is buttered all over. Put about ¾ of the mincemeat in the baking pan and spread it around evenly.

Melt the remaining butter in a small pan. Add the milk and eggs and mix well.

In another bowl, mix the remaining dry ingredients. Add the wet ingredients and mix until smooth and satiny.

Pour ½ of the batter over the mincemeat, and spread the remaining mincemeat evenly on the batter. Pour on the remaining batter. Bake for 35 minutes or until the cake tests done. Cool the cake and invert onto a plate. Serve with whipped cream.

Green Tomato Custard Pie

Yield: 8 servings
Time: 1 hour, plus chilling time

Crust:

1 cup unbleached flour
¼ teaspoon salt
⅓ cup vegetable shortening or butter
2–4 tablespoons cold water

Filling:

5 eggs
½ cup sugar
¼ cup maple syrup
2 teaspoons cider vinegar
½ teaspoon salt
½ teaspoon cinnamon
¾ cup Green Tomato Puree (p. 16)
2 cups milk, heated
¼ cup heavy cream

Combine the flour and salt. Cut the shortening into the flour and blend until the mixture resembles small peas. Add the water and mix until the dough forms a ball. Refrigerate the dough for ½ hour.

Preheat the oven to 425 degrees F.

Roll out the pie crust on a lightly floured board. Fit the crust into a 10-inch pie pan and flute the edges. Prick the crust with a fork. Bake for 15 minutes. Remove the crust from the oven, and reduce the oven temperature to 350 degrees F.

Beat together the eggs, sugar, maple syrup, vinegar, salt, and cinnamon. Stirring constantly, slowly add the tomato puree and heated milk, alternately.

Pour the custard mixture into the baked pie crust and place in the oven. Bake for 30 minutes, or until the custard is set. Remove from the oven and chill.

Beat the heavy cream until soft peaks are formed, spread the cream over the pie, and serve.

A custard is cooked when a knife inserted into the custard comes out clean.

Surprising Brown Betty

Yield: **6 generous servings**

Time: **1½ hours**

2 tablespoons butter

5 large green tomatoes, chopped

1 teaspoon cinnamon

1 teaspoon allspice

1 teaspoon nutmeg

½ cup raisins

1 lemon

1 cup brown sugar

½ cup butter

1 cup bread crumbs

1 cup wheat germ

1 cup heavy cream, whipped

Preheat the oven to 350 degrees F.

Melt 2 tablespoons of butter in a large saucepan, and add the tomatoes, spices, and raisins. Grate the peel of the lemon and add the peel to the tomatoes. Stir in the brown sugar.

In a separate pan, melt ½ cup of butter and add the bread crumbs and wheat germ to it.

Butter a covered casserole dish and put in ½ of the crumb mixture. Add the tomato mixture and squeeze the juice of the lemon over it. Pour on the remaining crumb mixture. Cover the casserole and bake for 40 minutes. Remove the cover and bake for 10 minutes more. Serve with whipped cream.

Sally's Green Tomato Fruit Cake

Yield: 10 servings
Time: 1½ hours

2½ cups chopped green tomatoes
1 cup water
½ cup golden raisins
½ cup black raisins
¾ cup chopped walnuts
½ cup butter
½ teaspoon nutmeg
½ teaspoon allspice
2 teaspoons cinnamon
1½ cups brown sugar, firmly packed
1 teaspoon vanilla
2 cups unbleached flour, sifted
5 teaspoons baking powder
¼ cup confectioners' sugar
1½ cups heavy cream, whipped

Preheat the oven to 350 degrees F.

In a medium-size saucepan, combine the tomatoes, water, raisins, walnuts, butter, spices, and brown sugar. Heat to boiling and simmer for 4 minutes. Cool to lukewarm. Add the vanilla.

Sift together the flour and baking powder, and add to the tomato mixture. Beat well. Pour into an 8-inch by 10-inch greased pan. Bake for 30–35 minutes or until the cake tests done. Cool completely.

Dust with confectioners' sugar. Serve with whipped cream.

> A quick bread or cake is done when the bread or cake has pulled away from the side of the pan slightly, and it bounces back when pressed lightly on the top.

Fruited Tea Bread

Yield: 12 servings

Time: 1 hour

1 orange

¼ cup melted butter

¾ cup brown sugar

2 eggs

½ teaspoon lemon extract

1 cup finely chopped green tomatoes, with juice

2½ cups unbleached flour, sifted

1 tablespoon baking powder

½ teaspoon baking soda

½ teaspoon salt

½ cup candied fruit

½ cup walnuts

1 cup heavy cream, whipped (optional)

Preheat the oven to 350 degrees F.

Grate the orange rind and set aside. Extract 3 tablespoons of juice from the orange. Set aside.

Cream together the butter, brown sugar, and eggs. Add the lemon extract, tomatoes, and orange juice; mix.

In a separate bowl, mix all the dry ingredients and add them alternately with the candied fruit, grated orange rind, and nuts to the tomato mixture. Stir until well blended.

Grease a bundt pan or a ring-shaped cake pan and pour in the batter. Bake the cake for 40 minutes, or until it tests done. Allow the cake to cool for 10 minutes and then unmold. Serve with whipped cream, if desired.

Holiday Spice Bread

Yield: 3 loaves
Time: 1½ hours

5 cups unbleached flour, sifted
5 teaspoons baking powder
2 teaspoons baking soda
1 teaspoon salt
2 cups brown sugar, firmly packed
4 tablespoons cinnamon
2 teaspoons nutmeg
1 teaspoon ginger
1 teaspoon allspice
½ cup honey
1 cup vegetable oil
2 cups chopped green tomatoes
4 eggs
2 teaspoons vanilla
1 cup chopped walnuts
1½ cups raisins
2 tablespoons grated orange peel

Why 3 loaves of bread? Freeze some loaves to bring out for the holidays, for busy days, or for delicious food gifts.

Preheat the oven to 350 degrees F.
Sift together all the dry ingredients. In a separate bowl, mix the remaining ingredients and add to the dry ingredients. Beat well. Grease 3 loaf pans and divide the batter among them. Bake for 45 minutes, or until the cake tests done.

Sweet Green Tomato Muffins

Yield: 12 muffins
Time: 45 minutes

2 cups all-purpose or unbleached flour,
   sifted
1 tablespoon baking powder
½ teaspoon salt
2 teaspoons cinnamon
1 egg
¼ cup melted butter or vegetable oil
⅓ cup honey
1 cup milk
2 cups chopped green tomatoes
½ cup raisins

Preheat the oven to 450 degrees F.

Sift the flour, baking powder, salt, and cinnamon together.

In a separate bowl, beat the egg and combine it with the remaining ingredients. Make a well in the center of the dry ingredients, and pour the wet ingredients in with the dry ingredients all at once. Stir just enough to moisten the batter, about 15 strokes. The batter should look lumpy.

Grease 12 muffin cups or line with paper cup liners. Fill each cup ⅔ full of batter. Bake for 25 minutes or until well browned.

Whole Wheat Mincemeat Muffins

Yield: 12 muffins
Time: 45 minutes

1 cup unbleached flour, sifted
1 cup whole wheat flour, sifted
1 tablespoon baking powder
½ teaspoon salt
1 egg, beaten
¼ cup vegetable oil
1 cup milk
1 cup Green Tomato Mincemeat (p. 28)

Preheat the oven to 425 degrees F.

Sift all the dry ingredients into a large bowl. Mix all the wet ingredients in another bowl.

Make a well in the center of dry ingredients and pour in the wet ingredients. Mix just enough to moisten the flour. The batter should look lumpy. Grease 12 muffin cups or line with paper liners. Fill each cup ⅔ full. Bake for 25 minutes or until golden brown.

Cheesy Green Tomato Muffins

Yield: 12 muffins
Time: 45 minutes

2 cups unbleached flour, sifted
1 tablespoon baking powder
½ teaspoon salt
1 egg
¼ cup melted butter or vegetable oil
2 tablespoons honey
1 cup milk
2 cups chopped green tomatoes
1 cup grated cheddar cheese

Preheat the oven to 450 degrees F.

Sift the flour, baking powder, and salt together. Make a well in the center of the bowl. Beat the egg, and combine it with the remaining ingredients. Pour the wet ingredients in with the dry ingredients all at once. Stir just enough to moisten the batter, about 15 strokes. The batter should look lumpy.

Grease 12 muffin cups or line with paper cup liners. Fill each cup ⅔ full of batter. Bake for 25 minutes or until well browned.

Red Tomato Specialties

Appetizers/Soups/Dressings and Salads
Sauces/Side Dishes/Main Dishes
Breads and Desserts

Tomato Juice Drinks

Tomato juice is very adaptable. Here are some recipes of vegetable and herb combinations to enjoy.

Each drink is made with 1 cup of tomato juice combined with the other ingredients in a blender and served over ice. Use vegetable sticks—strips of green peppers, stalks of celery, carrot sticks, or scallions—as stirrers and garnishes.

New Orleans Sipper: Combine 1 strip green pepper, 1 teaspoon lemon juice, dash Worcestershire sauce, dash cayenne pepper, dash garlic powder, and 1 thick avocado slice.

South of the Border: Combine ½ teaspoon cumin, 2 teaspoons Red Taco Sauce, dash cayenne.

Summer Cooler: Combine 1 celery stalk, diced, ½ teaspoon mint, and crushed ice.

French Connection: Combine ½ cup diced peeled cucumber and ½ teaspoon tarragon.

Dilly Cukes: Combine ½ cup diced peeled cucumber, 1 large scallion, minced, ½ teaspoon dill, and crushed ice.

Here's to Your Health: Combine 1 stalk celery, 3 heaping tablespoons yogurt, 1 teaspoon lemon juice, dash Worcestershire sauce, and dash of Tabasco sauce.

Some Like It Hot: Mix 1 tablespoon prepared horseradish with 1 teaspoon lemon juice.

Sugar & Spice: Combine ½ teaspoon cinnamon, pinch fennel, and 1 teaspoon honey.

Orient Express: Combine 1 teaspoon tamari, dash 5-Spice Powder, and 1 scallion.

The Mediterranean: Combine ½ teaspoon basil, dash thyme, dash garlic powder, and 1 teaspoon lemon juice.

Bugs Bunny's Favorite: Combine 1 small diced carrot, 2 sprigs parsley, 1 stick green pepper, dash pepper, and dash Tabasco sauce.

The Texan's Hat: Combine ½ cup beef broth and dash spicy steak sauce.

Fisherman's Net: Combine ¼ cup clam juice and ½ stick celery, diced.

Russian Cocktail: Combine ½ teaspoon caraway seeds and dash fennel. Top with yogurt.

Rosy Cheese Wafers

Yield: 60 wafers

Time: 45 minutes, plus chilling time

¼ **pound butter or margarine, at room temperature**

½ **pound Brie or cheddar cheese, cubed**

1⅓ **cups unbleached flour**

½ **teaspoon Tabasco sauce**

¼ **teaspoon salt**

¼ **cup tomato paste**

¼ **cup poppy, caraway, or sesame seeds**

So colorful and delicious you will be tempted to double the recipe.

Cream together the butter and cheese in a food processor fitted with a metal blade, or with a mixer. Add the flour, Tabasco sauce, salt, and tomato paste. Mix together until a ball is formed. Divide the dough into 4 parts and form into logs, 1 inch in diameter. Wrap in wax paper and chill until firm.

Preheat the oven to 400 degrees F. Slice the logs into ¼-inch circles and arrange 1 inch apart on an ungreased baking sheet. Sprinkle seeds on each wafer. Bake for 8–10 minutes, or until the edges are golden. Serve immediately.

Super-Duper Nachos

Yield: 6 servings
Time: 20 minutes

12 corn tortillas
¾ cup Red Taco Sauce (p. 91)
1 avocado, diced
1 cup grated cheddar or Monterey Jack
   cheese

Preheat the oven to 400 degrees F.

Cut the tortillas into triangles and arrange them in a shallow baking dish. Spread the Taco Sauce over the tortillas, leaving a corner of each tortilla uncovered to make them easier to pick up. Sprinkle the avocado pieces and the grated cheese on top. Bake for 10 minutes. Broil for 5 minutes or until the cheese is bubbly brown. Serve hot.

Variation: Cooked ground beef sprinkled on top before the nachos are baked make this a hearty snack.

Chili Con Queso

Yield: 8 servings
Time: 15 minutes

2 tablespoons vegetable oil
1 medium-size onion, diced
2 red tomatoes, finely diced
8 ounces cream cheese
½ pound Monterey Jack cheese, grated
1 jalapeño pepper, finely diced
2 teaspoons chili powder

In a medium-size frying pan, heat the oil, and sauté the onion for 5 minutes. Add the tomatoes, and continue cooking for 5 more minutes. Turn the heat down low, and add the remaining ingredients. Cook, stirring constantly, until the cheeses are melted. Serve warm with taco chips.

Tomato Swiss Cheese Soup

Yield: 12 servings
Time: 1 hour

½ cup butter
¾ cup unbleached flour
4 cups milk
4 cups red tomato juice or puree
¾ pound Swiss cheese, grated
2 teaspoons thyme
1 teaspoon basil
salt and pepper to taste
herb croutons

Melt the butter in a large soup pot. Add the flour and stir, mashing any lumps with the back of your spoon. Cook for 2 minutes. Slowly add the milk a little at a time, stirring after each addition until smooth. Add the remaining ingredients, except the croutons, and turn the heat to low. Cook until the cheese is melted. Do not boil this soup, as it will curdle, and the cheese will form into tiny lumps. Garnish with herb croutons.

Tomato Corn Chowder

Yield: 10 servings
Time: ½ hour

½ cup butter or margarine
½ cup unbleached flour
4 cups milk
8 cups Stewed Red Tomatoes (p. 97) or
   8 cups canned tomatoes
4 cups whole kernel corn (fresh or frozen)
salt and pepper to taste

Melt the butter over medium heat in a large pot. Mix in the flour until you have a smooth paste. Slowly add the milk, a little at a time, and continue stirring to prevent any lumps from forming. Add the stewed tomatoes and the corn. Cook over low heat until the corn is heated through. Correct the seasoning. Serve hot and garnish each bowl with crumbled bacon, croutons, or minced black olives, if desired.

Cream of Tomato Soup With Wine

Yield: 12 servings
Time: 1 hour

4 tablespoons butter
1 large onion, diced
3 cloves garlic, minced
10 large red tomatoes, chopped
½ cup unbleached flour
2 cups chicken or vegetable broth
1 pint cream or milk
1 cup white vermouth or dry white wine
salt and pepper to taste
lemon juice to taste
1 tablespoon dried dill or
  1 sprig for each bowl

Melt the butter in a soup pot, and sauté the onion and garlic until they are translucent. Add the tomatoes to the pot. Cook over medium heat for 5 minutes. Sprinkle the flour over the tomatoes and stir it in well. Add the remaining ingredients, except the lemon juice and the dill, and cook for 30 minutes.

Process in a blender or food processor until smooth. Reheat gently and adjust the seasoning. Add the lemon juice to taste and garnish with dill.

> When onions are sautéed until tender, they appear translucent. The color is more clear than white, and not browned.

Tortilla Soup

Yield: 6–8 servings
Time: 1 hour

2 tablespoons vegetable oil
1 large onion, diced
5 stalks celery, diced
1 quart canned red tomatoes
2 cups water or chicken broth
salt and pepper to taste
dash Tabasco sauce
12 soft corn tortillas
½ pound Monterey Jack cheese, grated
2 ripe avocados, diced
1 cup sour cream

Preheat the oven to 350 degrees F.

Heat the oil, and sauté the onion and celery in a soup pot. Just as the onion becomes limp and translucent, add the tomatoes, water, salt, pepper, and Tabasco sauce. Break the tomatoes up with your spoon. Cover the pot and simmer gently for 45 minutes.

Stack the tortillas in 2 piles and cut them into long strips. Spread the strips out on a cookie sheet and bake for 20 minutes, or until crispy.

Just before you are ready to serve, put some of the cheese and avocados into each soup bowl. Stand the tortilla strips around the inside edge of each bowl so that they stand above the rims of the bowls. Ladle the hot soup over the cheese and avocados and garnish with a dollop of sour cream. Serve immediately.

Hungarian Cabbage Tomato Soup

Yield: 10 servings
Time: 1 hour

3 tablespoons vegetable oil or butter
1 large onion, diced
8 cups shredded green cabbage
   (1 medium-size cabbage)
½ cup unbleached flour
4 cups Stewed Red Tomatoes (p. 97) or
   4 cups chopped red tomatoes
2 cups red tomato juice or puree
1 tablespoon caraway seeds
2 cups chicken stock or water
2 tablespoons honey or sugar
juice of 1 lemon
salt and pepper to taste
1 cup sour cream (optional)

This soup is perfect for a crowd. The recipe can be doubled or tripled with ease.

In a large soup pot, heat the oil. Add the onion and cabbage, and cook until the cabbage is limp but not brown. Sprinkle the flour over the cabbage, and mix completely. Add the remaining ingredients, except the sour cream, and cover the pot. Simmer for at least ½ hour, or as long as possible. Taste and correct the seasoning. Pass a bowl of sour cream at the table for a garnish.

Variations: This soup is equally delicious when made without oil, if the cabbage and onion are cooked in a little water first to wilt them.

When 1 cup of diced beets is added with the tomatoes, the color of the soup changes from red to deep purple, and you have a wonderful borscht.

Italian Fish Soup

Yield: 4–6 servings
Time: 45 minutes

2 tablespoons olive oil
1 medium-size onion, diced
1 medium-size zucchini, diced
1 large green pepper, diced
1 large clove garlic, minced
4 cups red tomato juice or puree
¼ cup fresh parsley
1 teaspoon basil
½ teaspoon oregano
1 teaspoon thyme
½ cup red or white wine
1 cup cooked rice or ¼ cup raw rice
2 cups water
½ pound boneless fish, cut in pieces, or
  1 cup minced clams, or ½ pound shrimp
salt and pepper to taste
Parmesan cheese

This soup is both elegant and simple to make. It's perfect for company.

In a medium-size soup pot, heat the oil, and sauté the onion, zucchini, pepper, and garlic for 10 minutes. Add the tomato puree, herbs, wine, rice, and water. Simmer for at least 15 minutes. Add the fish and cook for at least 10 minutes over low heat. This soup improves in flavor the longer you cook it. Correct the seasoning, and serve garnished with Parmesan cheese.

Variation: Small meatballs made from bulk sausage can be simmered in water until done, and then added to the soup, instead of the fish.

Red Gazpacho

Yield: 6–8 servings
Time: ½ hour, plus chilling time

4 large red tomatoes, quartered
2 medium-size cucumbers, peeled and
   coarsely chopped
1 medium-size onion, coarsely chopped
1 green pepper, coarsely chopped
2 stalks celery, coarsely chopped
3 cups tomato juice
1 large clove garlic, minced
¼ cup red wine vinegar
2 red tomatoes, finely diced
½ cucumber, finely diced
1 green pepper, finely diced
4 scallions, finely diced
salt and pepper to taste
Tabasco sauce to taste
1 cup herb croutons

Combine the coarsely chopped vegetables in a blender with a little of the tomato juice. Blend until smooth.

Pour the blended vegetables into a large bowl. Stir in the remaining tomato juice and vinegar. Add the finely diced vegetables. Add salt and pepper and Tabasco sauce to taste.

To make the soup thinner, add water and adjust the vinegar and seasonings. Serve icy cold, garnished with herb croutons.

Speckled White Gazpacho

Yield: 6–8 servings
Time: ½ hour, plus chilling time

4 medium-size cucumbers, peeled and coarsely chopped
1 small onion, diced
1 large clove garlic, minced
1 quart buttermilk
2 cups sour cream
2 tablespoons lemon juice
2 teaspoons dill
4 red tomatoes, finely diced
1 bunch parsley, finely chopped
1 green pepper, finely diced
½ cucumber, finely diced
salt and pepper to taste

Combine the coarsely chopped cucumber, onion, garlic, and buttermilk in a blender and process until smooth. Pour into a bowl, and add the remaining ingredients. Serve icy cold.

> Buttermilk is an excellent substitute for sour cream. It adds the same creamy texture to a soup, but it has far fewer calories.

Rosy Potato Leek Soup

Yield: 8 servings
Time: 1 hour

2 tablespoons vegetable oil
4 leeks, sliced (white part only)
4 medium-size potatoes, diced
4 medium-size red tomatoes, cut
  in chunks
5 cups rich chicken or vegetable broth
2 teaspoons dill
1 cup heavy cream
salt and pepper to taste

This soup can be served hot or cold.

In a medium-size soup pot, heat the oil, and sauté the leeks over medium heat for 10 minutes. Add the potatoes, tomatoes, broth, and dill. Cover the pot and simmer for 30 minutes. Cool the soup slightly and puree small amounts in a blender until smooth. Reheat gently and add the cream. Add the salt and pepper to taste. Serve garnished with dill.

Vegetable broth can be substituted for any of the recipes that call for beef or chicken broth. Tamari, a type of soy sauce, is a good flavor enhancer in broths.

Tomato Vinaigrette

Yield: 1¾ cups
Time: 5 minutes

1 red tomato, peeled and quartered
1 cup salad oil
¼ cup wine vinegar
¼ cup lemon juice
1 tablespoon Dijon-style mustard
½ teaspoon basil
2 cloves garlic, minced
dash salt and pepper

Combine everything in a blender or food processor and process until well blended.

Vermonter Sweet and Sour Dressing

Yield: 2½ cups
Time: 10 minutes

1 cup vegetable oil
1 cup tomato catsup
1 cup wine vinegar
½ cup maple syrup or honey
1 teaspoon chili powder
1 teaspoon Worcestershire sauce
¼ teaspoon garlic powder

Mix all ingredients well in a jar or bowl with a whisk. Refrigerate.

Variation: To make a low-calorie version of this dressing, eliminate the oil and use half the amount of vinegar, half the maple syrup, and half the seasonings. This oil-less version also makes a tasty marinade for beef, lamb, and chicken.

Tangy Tomato Dressing

Yield: about 3½ cups
Time: 5 minutes

½ lime
2 cups vegetable or salad oil
1 cup wine vinegar
½ cup tomato paste
1 tablespoon Dijon-style mustard
2 tablespoons fresh dill or 2 teaspoons
   dried dill
salt and pepper to taste
1 tablespoon lemon juice

Chop the lime, including the peel, into small pieces, and combine all the ingredients in a blender or food processor. Blend until smooth and light orange in color. Serve on any salad.

For a new approach to sliced tomatoes, cut the tomato from the stem end to the bottom, the opposite direction from a "sandwich cut." Arrange the sliced tomatoes and slices of cucumber or onion in a salad.

Rosy Cream Salad Dressing

Yield: 2 cups
Time: 10 minutes, plus chilling time

1 tablespoon sugar or honey
2 teaspoons Dijon-style mustard
1 teaspoon salt
dash cayenne pepper
1 cup salad oil
1 egg
1 cup red tomato pulp or puree
3 tablespoons cornstarch
¼ cup wine vinegar
1 teaspoon dill

The dressing also makes an excellent vegetable dip.

Combine the sugar, mustard, salt, pepper, oil, and egg in a mixing bowl, and set aside.

Pour the tomato pulp and cornstarch into a blender and mix until smooth. Heat this mixture over medium heat, stirring constantly, until the sauce thickens.

Pour the hot mixture into the other ingredients and beat vigorously, until the dressing is smooth. Chill until ready to use.

For best flavor, fresh salad tomatoes should be served at room temperature.

Presto Pesto Salad

Yield: 6–8 servings
Time: 20 minutes

Salad:
4 large red tomatoes
12 cups raw spinach
5 scallions, finely diced
1½ cups cucumber pieces
½ cup walnut pieces
½ cup Parmesan cheese

Dressing:
1 cup olive oil
½ cup lemon juice
3 cloves garlic, minced
3 teaspoons dried basil or 3 tablespoons fresh basil
salt and pepper to taste

Fill a salad bowl with washed spinach leaves that have been torn into bite-size pieces. Slice the tomatoes and quarter the slices. Arrange the tomatoes on top of the spinach. Sprinkle the scallions, cucumbers, Parmesan cheese, and nuts on top.

Put the oil, lemon juice, garlic, basil, salt, and pepper in a blender and blend well for 2 minutes. Pour the dressing into a pitcher and serve with the salad.

Tomato Feta Salad

Yield: 8 salad servings
Time: 10 minutes, plus marinating time

Salad:

6 medium-size red tomatoes, sliced
1 medium-size cucumber, sliced
1 small red onion, sliced and separated
   into rings
parsley to garnish
¼ pound feta cheese, crumbled

Dressing:

½ cup olive oil
⅓ cup red wine vinegar
2 sprigs parsley, without the stems
½ teaspoon basil
¼ teaspoon salt
¼ teaspoon black pepper
1 clove garlic, minced
1 teaspoon honey or sugar

Arrange the slices of tomato and cucumber on a lipped serving platter. Arrange the onion rings and feta cheese on top.

Put the dressing ingredients in a blender or food processor and process until well blended. Pour over the salad.

Let the salad marinate for at least 45 minutes before serving. Decorate with parsley.

To substitute fresh herbs for dried in a recipe, use 1 tablespoon minced fresh herbs for each ½ teaspoon of dried herbs the recipe called for. In dishes served hot, add fresh herbs during the last 15 minutes of cooking for best results; long cooking of fresh herbs reduces their flavors. In dips, chilled soups, and other cold foods, add the herbs several hours ahead, so the flavors can blend. In salad dressings, however, fresh herbs have the best color and flavor if added no more than 15 minutes before serving time.

Tabouli

Yield: 6–8 servings
Time: 1 hour

Salad:
1½ cups water
1½ cups bulgur
2 cups parsley, minced
1 cucumber, diced
4 red tomatoes, diced
1 green pepper, diced
1 bunch scallions, diced
½ pound feta cheese, crumbled
1 cup Greek black olives

Dressing:
¼ cup olive oil
½ cup lemon juice
1 clove garlic, minced
½ cup fresh mint or 3 tablespoons
   dried mint
½ teaspoon salt
¼ teaspoon pepper

This salad is especially good for picnics because it stores and keeps well.

In a 1-quart saucepan, bring the water to a boil. Turn off the heat and add the bulgur. Cover and let sit for 15 minutes.

In the meantime, combine all the other salad ingredients in a serving dish. Combine the dressing ingredients in a blender, blend for 1 minute, and set aside.

Uncover the pot and fluff the bulgur with a fork. Let sit until cool. Then mix the bulgur with the vegetables. Pour the dressing over everything and allow to sit for 15 minutes before serving.

Middle Eastern Picnic Salad

Yield: 8 servings
Time: ½ hour, plus marinating time

Salad:

6 medium-size red tomatoes, chopped
2 green peppers, diced
1 small red onion, diced
1 small unpeeled cucumber, diced
½ cup chopped fresh parsley
½ cup sliced ripe black olives
¼ pound feta cheese, crumbled
2 cups cooked, cubed chicken or turkey
   (optional)

Dressing:

1 cup olive oil
¼ cup lemon juice
¼ cup wine vinegar
1 tablespoon Dijon-style mustard
dash salt and pepper
2 cloves garlic, minced

This salad can be served on a bed of lettuce or stuffed in pita bread to make sandwiches.

Combine all the salad ingredients in a large serving bowl. Combine all the dressing ingredients in the blender, and mix well. Pour over the salad. Let the salad marinate for at least ½ hour.

Use 3 times the amount of prepared mustard for dried mustard—unless the prepared mustard is very sharp, then use only 2 times as much.

Red Tomato Dressings and Salads/87

Antipasto Salad

Yield: 6–8 servings
Time: 1 hour

Salad:
2 cups red cherry tomatoes, cut in halves
1 cup pepperoni, diced
2 cups diced, cooked ham
2 cups diced provolone cheese
2 stalks celery, diced on the diagonal
1½ cups cubed cucumbers
½ cup sliced black olives
½ cup sliced green olives
1 green pepper, chopped
10 pepperoncini peppers (hot Italian peppers), left whole
1 bunch of scallions, diced
1 jar marinated artichoke hearts (optional)
lettuce to garnish

Dressing:
1 cup olive oil
½ cup wine vinegar
dash salt and pepper
1 tablespoon minced parsley
2 teaspoons Dijon-style mustard
1 clove garlic, minced

This salad makes a wonderful meal when served with hot French bread and followed by a rich dessert.

Mix all the salad ingredients, except the lettuce, in a bowl. Combine all the dressing ingredients in a blender and blend until smooth. Pour the dressing over the salad and marinate for at least ½ hour. Serve on a bed of lettuce.

Tomato Aspic Supreme

Yield: 8–10 servings

Time: 20 minutes, plus at least 2 hours cooling time

4 cups tomato juice or puree
⅓ cup wine vinegar
3 teaspoons Worcestershire sauce
dash Tabasco sauce or cayenne pepper to taste
salt and pepper to taste
2 cups sour cream
⅓ cup water
2 tablespoons gelatin
1½ cups diced vegetables (carrots, celery, green peppers)
salad greens or black olives

Using a wire whisk, mix the tomato juice, vinegar, Worcestershire sauce, Tabasco sauce, salt, pepper, and sour cream until smooth.

Place the water in a small saucepan and sprinkle the gelatin on the water and let it sit for 2 minutes. Heat the gelatin over low heat, stirring constantly. When the gelatin is dissolved, pour it into the tomato mixture and mix it in well.

Put the diced vegetables into a 6-cup mold and pour the tomato mixture on top. Chill in the refrigerator until set—at least 2 hours. When the mold is set, unmold it onto a serving plate. Decorate with salad greens around it and sliced olives on top.

To unmold a gelatin salad, first wet your serving plate with cold water. Then, run warm (not hot) water into a large bowl or in the sink. Dip the mold into the water up to the depth of the gelatin for about 15 seconds. Loosen the gelatin around the edge of the mold with the tip of a paring knife. Place the serving dish on top of the mold, and carefully turn it over. Gently shake the mold until the gelatin comes loose. Lift off the mold. If the gelatin doesn't come out readily, repeat the process.

Tomato Aspic With 34 Variations

Yield: 6 servings
Time: 20 minutes, plus chilling time

2 tablespoons plain gelatin
4 cups red tomato juice or puree
2 tablespoons lemon juice
salt and pepper to taste
dash Worcestershire sauce
dash Tabasco sauce
1–2 cups chopped foods

Aspic can be partially frozen and served as a refreshing first course.

Soak the gelatin in ½ cup of the tomato juice for 5 minutes. Meanwhile, heat the remaining tomato juice, lemon juice, Worcestershire, Tabasco, salt, and pepper. Add to the softened gelatin and mix well. Add 1–2 cups of chopped foods. Pour the mixture into an oiled mold and chill until firm—about 2 hours. Unmold and serve.

Tomato aspic can also be chilled in a large flat baking dish. When the aspic is firm, unmold it in a single sheet. Cut the aspic in half lengthwise. Place a sheet of aspic on a serving platter and top with potato salad. Slip the other sheet of aspic on top to make a lovely buffet sandwich salad.

Variations: Add any of these foods to the basic recipe.

Parboiled asparagus	Zucchini cubes	Mushroom slices	Baked roast beef
Cucumber chunks	Parboiled cauliflower	Green or black olives	Crumbled bacon
Celery slices	Parboiled broccoli florets	1 teaspoon basil	Sliced hard-boiled eggs
Diced green pepper	Parboiled peas	1 teaspoon dill	Cooked shrimp
Grated carrot	Cooked chick peas	1 teaspoon mint	Cooked oysters
Sliced radishes	Parboiled lima beans	2 tablespoons parsley	Cooked white fish
Parboiled corn	Scallion slices	Cream cheese balls	Tuna
Red onion slices	Salmon	Baked ham	
Sprouts	Cubes of cheese	Baked turkey	

Red Taco Sauce

Yield: **4 cups**
Time: **45 minutes**

2 tablespoons vegetable oil
1 medium-size onion, diced
1 green pepper, diced
2 stalks celery, diced
3 cloves garlic, minced
4 cups red tomato juice or puree
2 tablespoons chili powder
½ teaspoon ground cumin
cayenne pepper to taste
1 jalapeño pepper, finely diced (optional)

Serve this sauce with nachos, chicken, or any of the Mexican dishes in chapter 4.

Heat the oil in a medium-size saucepan. Sauté the onion, green pepper, celery, and garlic for 10 minutes. Add the remaining ingredients, season to taste, and simmer for at least 15 minutes.

Cocktail Sauce

Yield: 2 cups
Time: 10 minutes

1½ cups red tomato puree
4 heaping tablespoons prepared
 horseradish
2 tablespoons lemon juice
dash Tabasco sauce
salt and pepper to taste

Mix all the ingredients and let the sauce sit for at least 30 minutes. Serve with shrimp and other seafood.

Sweet and Sour Sauce

Yield: 2½ cups
Time: 20 minutes

1 cup red tomato juice or puree
1 cup cider vinegar
¼ cup lemon juice
1½ cups honey
1 tablespoon dried mustard
dash Tabasco sauce
½ green pepper, minced
salt to taste
2 tablespoons cornstarch
2 tablespoons water

This sauce is excellent with stir-fried beef, pork, chicken, fish, and chicken wing appetizers.

In a medium-size saucepan, combine all the ingredients, except the water and cornstarch. Cook for 10 minutes. Stir the water and the cornstarch together until smooth and add to the tomato sauce. Cook until thick and clear.

Salsa

Yield: 1½ cups
Time: 10 minutes

2 ripe red tomatoes, finely chopped
½ cup green peppers, finely chopped
1 sereno (hot) pepper, very finely diced
1 small onion, finely diced
2 cloves garlic, minced
juice from 1 lemon
salt and pepper to taste

This wonderful sauce is served with every meal in Mexico. It is great as a dip with corn chips or as a taco sauce.

Mix all ingredients and serve.

Celery Tomato Sauce

Yield: 5 cups
Time: 45 minutes

2 tablespoons vegetable oil
1 medium-size onion, finely diced
2 cups celery, finely diced
4 cups red tomato juice or puree
2½ teaspoons cilantro or fresh coriander
  leaves
1 teaspoon dill
salt and pepper to taste

This sauce goes nicely with meat loaf and other beef dishes.

In a medium-size saucepan, heat the oil and sauté the onion and celery for 5 minutes. Add the remaining ingredients and simmer for ½ hour.

Indian Curry Sauce

Yield: 4 cups
Time: 20 minutes

2 tablespoons vegetable oil
1 medium-size onion, diced
1 large stalk celery, diced
4 cups red tomato juice or puree
2 tablespoons honey
3–4 tablespoons curry powder

A wonderful spicy sauce to serve over steamed vegetables, fish, or chicken.

In a medium-size saucepan, heat the oil, and sauté the onion and celery for 10 minutes or until tender. Combine all the ingredients in a blender and blend for 30 seconds. Pour the sauce back into the pan and keep it warm.

Serve the sauce over steamed vegetables, such as cauliflower, brussels sprouts, and chickpeas.

Italian Sauce

Yield: 4 cups
Time: 45 minutes or more

2 tablespoons olive oil
1 small onion, diced
2 cloves garlic, minced
½ teaspoon thyme
½ teaspoon oregano
½ teaspoon basil
1 bay leaf
¼ cup minced fresh parsley
¼ cup Parmesan cheese
½ cup dry red wine
1 teaspoon sugar or honey
4 cups red tomato juice or puree
salt and pepper to taste

In a medium-size saucepan, heat the olive oil and sauté the onion and garlic for 5 minutes. Add the herbs and continue cooking for 2 minutes. Add the remaining ingredients and cook for at least 30 minutes. This sauce gets better the longer you cook it!

Variation: To make an Italian meat sauce, sauté 1 pound of ground beef with the onions and garlic. Drain off fat, and continue cooking as the recipe specifies.

Andrea's Rosy Basil Pasta Sauce

Yield: 4 servings
Time: 20 minutes

¼ cup butter
2 cloves garlic, minced
5 medium-size red tomatoes, chopped
1 cup coarsely chopped fresh basil leaves
¼ cup unbleached white flour
salt and pepper to taste
¼ cup milk or cream (optional)

This elegant, rich-tasting sauce can be whipped up in less time than it takes to boil the water to cook the pasta.

Melt the butter in a saucepan. Add the garlic and sauté for 1 minute. Add the tomatoes and the basil, and sauté for another 5 minutes. Pour the tomato mixture into a blender. Add the flour and blend until the mixture is smooth. Return the sauce to the pan. Season to taste. Thin with milk or cream, if desired. Cover the pot and cook the sauce on low heat for 10 minutes.

Serve this sauce over hot pasta with Parmesan or Romano cheese to sprinkle on top.

Stewed Tomatoes

Yield: 6 servings
Time: 20 minutes

1 tablespoon vegetable oil
10 large red tomatoes, cut in chunks
1 teaspoon honey
salt and pepper to taste
2 teaspoons cornstarch (optional)
2 tablespoons water (optional)

Heat the oil in a saucepan, add the tomatoes and honey, and simmer gently for 15 minutes, or until the tomatoes are soft. Season with salt and pepper. To thicken the tomatoes, combine the cornstarch with the water. Add to the tomatoes and cook until the sauce thickens.

Variations: This dish lends itself well to variations. Here are some additions to the basic recipe.

Onions: Sauté 1 small or medium-size diced onion in the oil for 5 minutes. Then add the tomatoes.

Garlic: Add 1 minced medium-size clove garlic to the tomatoes.

Herbs: Add 1½ teaspoons of either basil, thyme, cumin, marjoram, tarragon, or curry powder to the tomatoes.

Bread crumbs: Stir in ½–¾ cup cracker crumbs or wheat germ during the last few minutes of cooking time.

Cinnamon: Season with ¼–½ teaspoon cinnamon and salt and pepper to taste.

Cheese: Pour the stewed tomatoes into a casserole dish. Sprinkle the top with grated Parmesan cheese or grated cheddar and broil for a few minutes until bubbly.

Bacon: Sauté a few pieces of bacon until crisp, and pour off all but 2 tablespoons bacon grease. Sauté the tomatoes in the grease. Crumble the bacon into the tomatoes after you cook them.

Ham: Sauté cubes of ham with the tomatoes.

Broiled Tomatoes

Time: 10 minutes

1 medium-size red tomato per person
1 teaspoon melted butter or vegetable oil
   per person

Select 1 tomato per person. Cut it in half horizontally. Brush with melted butter. Broil the tomatoes on a low rack until the tops are browned but not burned and the tomatoes are heated all the way through.

Variations: Try one of these variations to make your dinner more exciting.

Herbs: Sprinkle 1 teaspoon of one of the following on each tomato before broiling: garlic, thyme, marjoram, oregano, rosemary, coriander, parsley, dill, chives, cumin, basil, chervil, tarragon, chili powder, or curry powder.

Breadings: Sprinkle bread crumbs or wheat germ on top of the tomatoes before broiling. Or combine equal amounts of breading and cheese and sprinkle on top before broiling.

Cheeses: Sprinkle on top of each tomato half before broiling: Parmesan, cheddar, Swiss, Blue cheese, feta, cream cheese, or any other cheese you have on hand.

Garnishes: Garnish before broiling with diced onion, crumbled bacon, sesame seeds, scallions, or anchovies.

Sour Cream Topping: Broil the tomatoes until they are heated. Combine 2 tablespoons sour cream and 2 teaspoons prepared mustard. Spread the sour cream mixture on top of the tomatoes and broil until the new topping is bubbling hot.

Deviled Tomatoes

Yield: 4 servings
Time: 20 minutes

4 large red tomatoes, halved
salt and pepper to taste
cayenne pepper to taste
½ cup bread crumbs
2 tablespoons butter
1 teaspoon Dijon-style mustard
¼ teaspoon Tabasco sauce
1 tablespoon Worcestershire sauce
2 teaspoons sugar
3 tablespoons cider vinegar
2 tablespoons unbleached flour

A spicy variation on grilled tomatoes. Serve this dish with a pasta or spinach salad for a light summer dinner.

Arrange the tomato halves in a shallow baking dish cut side up. Lightly sprinkle the tomatoes with salt, pepper, and cayenne pepper. Sprinkle the bread crumbs evenly over them.

In a small saucepan, combine the remaining ingredients, except the flour, and stir over medium heat until the butter is melted. Sprinkle the flour over the sauce and stir until blended.

Spoon the sauce on top of the tomatoes. Broil for 10 minutes, or until the crust is browned and bubbling.

Herbed Cherry Tomatoes

Yield: 6 servings
Time: 5 minutes

2–3 tablespoons butter
½ teaspoon basil
½ teaspoon thyme
1 clove garlic, minced
4 cups red cherry tomatoes
salt and pepper to taste

Melt the butter in a large frying pan, and sauté the herbs, garlic, and tomatoes for 2–3 minutes, or until hot. Add salt and pepper. Serve immediately.

Variation: Sauté 2 tablespoons parsley for 5 minutes; then add the tomatoes and salt and pepper.

Tomatoes and Okra

Yield: 6 servings
Time: 30 minutes

2 tablespoons vegetable oil
1 medium-size onion, diced
2 cloves garlic, minced
½ pound (4 cups) okra, diced
3 red tomatoes, diced
1 tablespoon cilantro or coriander leaves
1 teaspoon thyme

In a large frying pan, heat the oil, and sauté the onions and garlic until they appear translucent. Add the okra and cook for 5 minutes, stirring often. Add the remaining ingredients and turn the heat down. Cook for 15 minutes, or until the okra is tender.

Chili Corn Fry-up

Yield: 6–8 servings
Time: 30 minutes

2 tablespoons vegetable oil
1 clove garlic, minced
1 medium-size onion, diced
1 sweet red pepper, diced
1 green pepper, diced
2 cups whole kernel corn (fresh or frozen)
4 red tomatoes, diced
2 tablespoons chili powder
salt and pepper to taste

In a large frying pan, heat the oil, and sauté the onions and garlic for 5 minutes. Add the peppers and continue cooking for 5 minutes. Add the remaining ingredients and cook uncovered until the tomatoes are soft—about 5-10 minutes. Serve hot.

Lebanese Green Beans

Yield: 8 servings
Time: 25 minutes

2 tablespoons vegetable oil
1 medium-size onion, diced
2 cloves garlic, minced
6 medium-size red tomatoes, diced
4 cups green beans, diced
2 teaspoons cinnamon
½ teaspoon allspice
salt and pepper to taste

In a large frying pan, heat the oil, and sauté the onion and garlic for 5 minutes. Add the remaining ingredients. Cook until the green beans are tender.

Little Eggplant Pizzas

Yield: 6–8 servings
Time: 20 minutes

2 medium-size eggplants
½ cup olive oil
1 small onion, diced
2 cloves garlic, minced
4 cups red tomato juice or puree
½ teaspoon oregano
1 teaspoon basil
dash thyme
2 tablespoons minced fresh parsley
pinch rosemary, crumbled
1 teaspoon honey
salt and pepper to taste
1 cup Parmesan cheese

Slice the unpeeled eggplants into ½-inch circles and place the pieces on a baking sheet.

Brush each slice with olive oil and broil until the eggplant is golden brown. Turn the slices over, brush with oil, and broil until golden.

In the meantime, heat 1 tablespoon of oil in a saucepan and sauté the onion and garlic. Add the tomato puree, herbs, honey, and salt and pepper. Simmer for 10–15 minutes.

Carefully spread the sauce on each eggplant slice. Sprinkle with Parmesan cheese, and broil for 5 minutes, until the cheese is bubbling hot and the eggplant is heated through.

North End Beans

Yield: 6 servings
Time: 1 hour

½ cup water
2 10-ounce packages of frozen lima beans
   (3–4 cups, cooked)
2 tablespoons vegetable oil
1 small onion, diced
2 cloves garlic, minced
1 teaspoon basil
1 teaspoon oregano
1 teaspoon thyme
2 cups red tomato juice or puree
½ pound provolone cheese, grated

Preheat the oven to 350 degrees F.

Boil the water in a small saucepan and add the frozen lima beans. Cook for 5–10 minutes, or until the beans are just defrosted; drain.

In a separate saucepan, heat the vegetable oil, and sauté the onion and garlic for 5 minutes. Add the basil, oregano, thyme, tomato puree, and lima beans. Pour this mixture into an oiled baking dish. Sprinkle the cheese on top. Bake for 35 minutes.

> To sauté means to stir foods until tender in a little oil or butter over medium high heat.

Quick Creamy Tomato Crunch

Yield: 6 servings
Time: 1 hour

2 tablespoons vegetable oil
1 large onion, sliced
4 stalks celery, diced
6 large red tomatoes, sliced
1 cup bread crumbs
3 teaspoons dried basil or 3 tablespoons
   fresh basil
salt and pepper to taste
2 cups sour cream

Preheat the oven to 350 degrees F.

In a small frying pan, heat the oil, and sauté the onion and celery for 10 minutes.

Butter a baking dish and arrange half of the tomato slices in the bottom of the dish. Sprinkle half of the bread crumbs on top. Then spread the onions and celery on top of the bread crumbs, and sprinkle with half of the basil and some salt and pepper. Cover with another layer of tomato slices. Sprinkle the rest of the herbs over that. Then spread the sour cream on top. Cover with the remaining bread crumbs. Bake for 30 minutes.

Savory Red Tomato Crisp

Yield: 6–8 servings
Time: 1½ hours

2 cups bread crumbs
½ cup butter, melted
1 tablespoon oregano
2 eggs, beaten
1 teaspoon salt
2 tablespoons butter
1 onion, chopped
1 clove garlic, minced
1 green pepper, chopped
1 small zucchini, chopped (2 cups)
6 large red tomatoes, chopped
1 tablespoon chili powder
salt and pepper to taste
1 pound cheddar cheese, grated

Preheat the oven to 350 degrees F.

Combine the bread crumbs, melted butter, oregano, eggs, and salt. Press half the mixture into the bottom of a 9-inch by 13-inch baking dish.

In a large frying pan, melt 2 tablespoons butter, and sauté the onion, garlic, green pepper, and zucchini until just tender. Add the chopped tomatoes and cook for 5 minutes. Drain off most of the liquid. (Save the liquid to use in a soup.) Add the chili powder and salt and pepper to the vegetables, and spread the vegetable mixture on the crust. Cover with the grated cheddar cheese. Sprinkle on the remaining topping. Bake for 45 minutes.

Calzones

Yield: 12 servings
Time: 2½ hours

3 tablespoons dried baker's yeast
2 cups warm water
1 tablespoon honey
1 teaspoon salt
5 cups unbleached flour
1 head broccoli, chopped and steamed
1 pound ricotta cheese
1 small onion, finely minced
¾ cup Parmesan cheese
salt and pepper to taste
4 cups Italian Tomato Sauce (p. 95)
12 slices mozzarella cheese (½ pound)
12 slices ham (½ pound) (optional)

Calzones are well worth the extra time to prepare. If you are cooking for a small group, make extra and freeze the leftovers for another meal.

In a medium-size mixing bowl, combine the yeast, water, and honey. Let sit for 5 minutes. Add salt and ½ of the flour and stir to form a smooth dough. Add the remaining flour and mix in well. Pour the dough out onto a floured bread board and knead for 10 minutes. Cover and put in a warm place for 1 hour.

In the meantime, combine the broccoli, ricotta cheese, onion, Parmesan cheese, and salt and pepper.

When the dough has risen, punch it down and divide into 12 pieces. Roll each piece into a 6-inch circle. Spoon ½ cup of the broccoli mixture into each circle. Top with a small amount of Italian Tomato Sauce, 1 slice of mozzarella cheese, and 1 slice of ham. Moisten the edge of the circle with water and fold the dough over to make a pocket. Crimp the edge.

Bake the calzones on a greased baking sheet for 20 minutes, or until brown. Serve with warm Italian Sauce to spoon over the calzones.

Tomato Zucchini Tart

Yield: 8 servings
Time: 1½ hours

Crust:

2 cups grated raw potato
1 small onion, finely diced
1 egg, lightly beaten
¼ cup unbleached flour
dash salt

Filling:

2 cups grated cheese (cheddar, Swiss,
   mozzarella, or provolone)
2 cups thinly sliced zucchini
3 large red tomatoes, sliced
1 teaspoon basil
1 teaspoon oregano
½ teaspoon garlic powder
salt and pepper to taste
1 small onion, diced

Delicious and very easy.

Preheat the oven to 350 degrees F.

After grating the potato, squeeze out as much of the juice as possible. Combine the potato with the onion, egg, flour, and salt. Mix well. Pat this mixture into a well-greased 9-inch pie pan. Bake in the oven for 30 minutes while you prepare the vegetables for the filling.

When the crust has browned around the edge a bit, remove it from the oven. Layer half of the cheese in the crust. Arrange the zucchini slices on top of the cheese, and put the tomato slices on next. Sprinkle on the basil, oregano, garlic powder, salt and pepper, and onion. Top with the rest of the cheese. Bake for 45 minutes, or until the top is a lovely golden brown.

Crepes

Yield: 6–8 servings (12–16 crepes)
Time: 45 minutes

2 eggs, beaten
1 cup milk
1 cup unbleached flour
dash of salt
1 tablespoon melted butter or
    vegetable oil

Mix all the ingredients together in a mixing bowl, and beat with a hand or electric beater until the batter is completely smooth. Let the batter sit for 30 minutes. This batter will be thin, similar to cream.

In the meantime, make your filling. (There are recipes on pp. 109 and 110.)

When it's time to cook the crepes, heat a small, nonstick frying pay or omelet pan and melt a teaspoon of butter in the pan. Spread the butter around to cover the whole bottom of the pan and part way up the sides. Pour a little less than a ¼ cup of batter into the pan and quickly tip the pan, moving it in a circular fashion to evenly spread the batter over the bottom and part of the sides of the pan. Cook the crepe on medium for about 2 minutes, then flip it over and continue cooking for 30 seconds longer. Remove the crepe to a plate and fill.

When you get very fast at making crepes you can have 2 crepe pans going at the same time.

Italian Crepes

Yield: 8 servings (2 crepes each)
Time: 1 hour, plus time to make the crepes and sauce

¼ cup vegetable oil
½ pound mushrooms, sliced
1 medium-size onion, diced
3 cloves garlic, minced
4 pounds fresh leaf spinach or
  2 pounds frozen spinach
2 cups ricotta cheese
1 pound mozzarella cheese
16 crepes (p. 108)
3 cups Italian Tomato Sauce (p. 95)

Preheat the oven to 350 degrees F.

In a frying pan, heat the oil, and sauté the mushrooms, onion, and garlic until limp. Transfer to a mixing bowl. Steam the spinach until limp. Drain well and add to the mixing bowl. Add the ricotta cheese and ½ of the mozzarella.

Put ½ cup of the cheese and spinach filling in each crepe. Roll each crepe and place in a baking pan. Spoon Italian Tomato Sauce over the filled crepes and sprinkle the remaining mozzarella on top. Heat the crepes for 30 minutes and serve.

One pound of de-stemmed leaf spinach equals 1½ cups cooked spinach.

Spanish Crepes

Yield: 6 servings (2 crepes each)

Time: 1½ hours, plus time to make the
       crepes

Cheese Sauce:

½ cup butter or margarine

½ cup unbleached flour

2 cups milk

1 pound cheddar cheese, shredded

1 tablespoon Dijon-style mustard

dash Worcestershire sauce

salt and pepper to taste

Crepes:

¼ cup vegetable oil

1 large onion, diced

1 clove garlic, minced

1 cup diced raw pork

2 cups diced cooked ham

5 large red tomatoes, diced

¼ cup unbleached flour

1 cup beef broth

salt and pepper to taste

12 crepes (p. 108)

Melt the butter in a saucepan over medium heat. Stir in ½ cup flour, making sure there are no lumps. Slowly add the milk, a little at a time, stirring after each addition to remove all lumps. Add the cheese, mustard, Worcestershire sauce, and salt and pepper. Continue cooking until the sauce is thick and smooth. Set aside.

Preheat the oven to 350 degrees F.

In a large frying pan, heat the oil. Sauté the onion and garlic until the onion appears translucent. Add the pork, and cook until lightly browned. Stir in the ham and the tomatoes.

Make a paste with ¼ cup flour and ½ cup beef broth. Stir into the ham and tomatoes. Add the remaining broth and correct the seasoning. Simmer until the mixture thickens. Cool slightly.

Spoon ½ cup filling into each crepe. Roll each crepe and place in a baking dish. Spoon the cheese sauce over the crepes. Bake for 20 minutes, or until bubbly hot. Serve at once.

Ratatouille

Yield: 8 generous servings
Time: 45 minutes

3 tablespoons olive oil
1 large onion, sliced (2 cups)
2 cloves garlic, minced
1 large, unpeeled eggplant, diced (8 cups)
1 medium-size zucchini, diced (4 cups)
2 medium-size green peppers, diced
 (1 cup)
5 medium-size red tomatoes, diced
 (4 cups)
1 teaspoon basil
1 teaspoon oregano
1 teaspoon thyme
1 teaspoon marjoram
1 cup tomato paste
salt and pepper to taste
1 cup grated cheese (cheddar, Parmesan,
 provolone)

A wonderful French vegetable stew with everything in it but the kitchen sink . . . This is a dish that freezes well. Leftovers can be thawed and served, or incorporated into sauces and soups.

In a large pot, heat the olive oil, and sauté the onion and garlic until they appear translucent, but not browned. Add the eggplant, and cook over medium heat for about 5 minutes, stirring frequently. Add the remaining vegetables and herbs. Continue cooking for an additional 10 or 15 minutes, stirring often. Add the tomato paste. If the sauce is too thick, add a bit of water. Season with salt and pepper. Cook for 5 minutes over medium heat. Serve immediately, passing grated cheese as a garnish.

Tomato Tuna Gumbo

Yield: 8 servings
Time: 1 hour

3 tablespoons olive oil
2 medium-size onions, diced (2 cups)
3 cloves garlic, minced
1 large, unpeeled eggplant, diced (8 cups)
4 cups diced okra (fresh or frozen)
2 medium-size green peppers, diced
12 red tomatoes, chopped
½ cup tomato paste
1 cup chicken broth or water
¼ cup minced parsley
1 bay leaf
1 teaspoon thyme
1 teaspoon basil
1½ teaspoons Worcestershire sauce
dash cayenne pepper
salt and pepper to taste
1 large 12-ounce can of tuna
3 cups cooked rice

Heat the oil in a large pot, and sauté the onion and garlic. Add all of the remaining vegetables and cook for 15 minutes. Add the tomato paste, broth, herbs, Worcestershire sauce, salt, pepper, cayenne, and tuna. Cover and cook the gumbo for another 20 minutes. Serve over cooked rice.

If you have some leftover tomato paste after preparing a meal, flash freeze it in small amounts to use later in gravies, sauces, or soups. Simply line a baking sheet with waxed paper, and put dollops of paste (about a tablespoon each) on the sheet. Freeze uncovered until firm, then transfer the frozen paste drops into little plastic bags and store in freezer.

Fancy Chicken Pepperon

Yield: 6 servings
Time: 1 hour

½ cup vegetable oil or butter
6 chicken breasts or chicken pieces
2 large onions, diced
2½ cups diced red and green peppers
2½ cups diced red tomatoes
2 cloves garlic, minced
1 teaspoon basil
salt and pepper to taste
½ pound bulk sausage
½ pound mushrooms, sliced

Preheat the oven to 350 degrees F.

Heat ¼ cup oil in a large frying pan, and brown the chicken on both sides. Then put the chicken in a baking pan. In a medium-size saucepan, heat the remaining oil, and sauté the onions and peppers. When they begin to get limp, add the tomatoes, garlic, basil, and salt and pepper. Cook for 15 minutes on medium heat, stirring constantly. Ladle the sauce over the chicken.

Brown the sausage, drain, and reserve 2 tablespoons of fat. Crumble the sausage over the chicken. In the reserved sausage fat, sauté the mushrooms. Sprinkle the mushrooms over the chicken. Cover the baking dish and bake for 30 minutes. Remove the cover for the last 15 minutes. Serve with rice.

Chicken Victoria

Yield: 6 servings
Time: 1 hour

Chicken Roll:

3 boneless, skinless chicken breasts, split
  in half
6 slices smoked ham (¼ pound)
6 slices Swiss cheese (¼ pound)
1 egg
½ cup milk
1½ cups bread crumbs
4 tablespoons vegetable oil

Sauce:

¼ cup butter
2 cloves garlic, minced
6 medium-size red tomatoes, chopped
¼ cup unbleached white flour
¼ cup half-and-half or heavy cream
salt and pepper to taste

Preheat the oven to 350 degrees F.

Put the chicken breasts between 2 pieces of waxed paper and pound them with a meat mallet until the breasts are ½ inch thick. Place a piece of ham and a piece of cheese on top of each chicken breast. Roll the breast up to make a log.

In a small bowl, beat the egg and milk together. Dredge the chicken roll in the egg mixture and then in the bread crumbs.

Heat the oil in a frying pan, and brown the chicken roll on both sides. If the chicken begins to unroll, hold it together with wooden toothpicks. (Remove the toothpicks before baking.)

Place the chicken rolls in a baking dish and bake for 30 minutes.

While the chicken bakes, make the sauce. Melt the butter, and sauté the garlic for 1 minute. Add the tomatoes and cook for 10 minutes. Transfer the mixture to a blender, add the flour, and blend until smooth. Return the mixture to the pan, add the cream, and season to taste. Keep warm, on low heat, until the chicken is baked.

To serve, pour the sauce into a serving dish and place the chicken breasts in the sauce.

Chicken Stir Fry

Yield: 4–5 servings
Time: 30 minutes

¼ cup soy sauce
¼ cup dry sherry
¼ cup water
2 tablespoons cornstarch
1 chicken breast, diced in 1-inch pieces
6 tablespoons vegetable oil
1 medium-size onion, sliced in rings
2 cloves garlic, minced
1 green pepper, coarsely diced
½ cup sliced water chestnuts
2 red tomatoes, cut in wedges

Mix the soy sauce, sherry, water, and cornstarch to make a marinade for the chicken. Put the chicken in the marinade and leave it for 15 minutes while you prepare the vegetables.

Heat 3 tablespoons of oil in a wok or a large frying pan. Add the onion and stir fry for 1 minute. Add the garlic, pepper, and water chestnuts, and continue stir-frying until the vegetables are tender crisp. Remove the vegetables from the pan.

Heat the remaining 3 tablespoons of oil in the wok or frying pan, and pour in the chicken and the marinade. Stir fry until the chicken is cooked—about 5 minutes. Add the vegetables and the tomatoes, and cook for about 2 minutes, or until the tomatoes are warm. Serve with rice and soy sauce.

> Stir-fry means to stir and toss briskly in oil over high heat.

Eggs New Orleans

Yield: 6 servings
Time: 1½ hours

New Orleans Sauce:

1 medium-size onion, chopped
2 tablespoons olive oil
1 medium-size, unpeeled eggplant, cubed
4 stalks celery, diced
2 green peppers, diced
1 medium-size zucchini, cubed
8 medium-size red tomatoes, diced
½ cup tomato paste or sauce (optional)
1 teaspoon basil
1 teaspoon oregano
1 teaspoon marjoram
1 teaspoon thyme
salt and pepper to taste

Rice:

4 cups water
1 teaspoon salt
2 cups uncooked brown rice
1 pound bulk sausage meat (optional)

This is one of my favorite Sunday brunch dishes. Sometimes I make the rice and the New Orleans Sauce in advance. When it is time to serve, I reheat the sauce and rice, and whip up the Egg and Hollandaise Sauce Topping.

To make the New Orleans sauce, sauté the onion in olive oil. Add the rest of the vegetables (and more oil if necessary). Cook over medium heat until the vegetables are soft—about 10 minutes. Add the tomato paste, the herbs, and salt and pepper. Continue cooking for about 1 hour over a low heat.

At this point, this sauce can be refrigerated or frozen until you are ready to serve it. Then just reheat it.

To make the rice bed, combine the water and salt, and bring the water to a boil. Add the rice, stir once with a fork, turn the water down to a simmer, and cover. Cook for 45 minutes, or until the water is absorbed, and the rice is soft, but not mushy.

While the rice is cooking, break up the sausage, and sauté it in a frying pan, until browned. Transfer the sausage to paper towels, and drain off the grease.

When the rice is cooked, mix the sausage and rice and keep the mixture warm. This mixture can be refrigerated or frozen until you are ready to serve it. To reheat, put the rice in a baking dish, cover, and warm in the oven at 350 degrees F. for 20 minutes.

Hollandaise Sauce and Egg Topping:
salted water for poaching the eggs
½ pound butter, melted
6 egg yolks
¼ cup lemon juice
¼ teaspoon salt
dash cayenne pepper
12 eggs

A few minutes before you are ready to serve this dish, make sure the sauce is heated and the rice is warm. Fill a large frying pan with salted water for poaching the eggs, and heat to boiling. While the water is heating, melt the butter for the Hollandaise Sauce, and keep it hot on the stove. Assemble the rest of the Hollandaise Sauce ingredients—egg yolks, lemon juice, salt, and cayenne—in a blender and set aside.

When the poaching water starts to boil, lower the water temperature to a simmer, gently crack the eggs, and slip them out of their shells and into the water. Poach until the white is set—3–5 minutes.

Begin to assemble the Eggs New Orleans. Spoon a bed of rice and sausage on each plate. Cover the rice with hot New Orleans Sauce. Carefully lift the poached eggs out of the water with a spatula, and place 2 poached eggs on top of the sauce.

Then make the Hollandaise Sauce. Turn the blender—which already has in it the egg yolks, lemon juice, salt, and cayenne—on high for 3 seconds. With the blender still on, pour the *hot* butter into the egg mixture in a steady stream. This should take about 30 seconds.

Pour ¼ cup of Hollandaise Sauce over each plate and serve immediately.

Many cooks tell you to lift poached eggs out of the water with a slotted spoon, but a spatula works better because there are no sharp, curved edges that can break the yolk.

Herbed Tomato Corn Bread

Yield: 8 servings
Time: 1 hour

1 cup unbleached white flour
⅔ cup whole wheat flour
3 teaspoons baking powder
½ teaspoon salt
2 tablespoons sugar or honey
¾ cup cornmeal
½ teaspoon basil
½ teaspoon thyme
1 teaspoon dill
1 egg
1½ cups red tomato juice or puree
¼ cup vegetable oil

This bread is such a lovely color that people rave about it as they gobble it up. Perfect for a barbecue.

Preheat the oven to 425 degrees F.

Sift the flours with the baking powder, salt, and sugar. (If you are using honey, mix it with the tomato juice.) Stir in the cornmeal and herbs. In another bowl, beat the egg. Stir in the tomato juice and oil, and add this mixture all at once to the dry ingredients. Stir just enough to moisten the mixture.

Pour the batter into a greased 8-inch by 8-inch pan and bake for 30 minutes, or until golden on the top.

> Whole wheat flour is a nice substitute for white flour. Use half whole wheat and half white flour.

Tomato Herb Bread

Yield: 2 loaves
Time: 4 hours

¼ cup warm water
1 tablespoon dried baker's yeast
2 tablespoons honey
2 cups warm red tomato juice or puree
2 tablespoons vegetable oil
2 teaspoons salt
2 teaspoons basil
1 teaspoon marjoram
4 cups unbleached flour
2 cups whole wheat flour
1 egg, beaten

Pour the water into a large mixing bowl. Stir in the yeast and the honey. Let this mixture sit for 10 minutes until it is foamy.

Add the tomato juice, oil, salt, and herbs. Stir in 3 cups of unbleached flour 1 cup at a time, and beat until smooth and elastic. Mix in the rest of the unbleached flour, and beat well. Sprinkle a large board with ½ cup of whole wheat flour, and turn the dough out onto the board. Knead in the remaining whole wheat flour. If the dough is too sticky, add unbleached flour until the dough is a firm, smooth consistency. Knead the dough for 10 minutes.

Oil a mixing bowl and place the dough in the bowl. Let the bread rise for 1½ hours in a warm place.

Punch the dough down and shape it into 2 loaves. Put the bread in greased loaf pans and let it rise for 45 minutes. Preheat the oven to 375 degrees F.

Brush the crusts with a beaten egg. Bake for 25–30 minutes.

Grandma's Surprise Cake

Yield: 8 servings
Time: 1 hour

2 cups unbleached flour
½ teaspoon salt
1½ teaspoons cinnamon
½ teaspoon nutmeg
1 teaspoon ginger
½ teaspoon ground cloves
1 teaspoon baking soda
2 tablespoons butter
1 cup sugar or ¾ cup honey and
    ¼ cup flour
¾ cup red tomato juice or puree
1 cup chopped walnuts
½ cup raisins
1½ cups Cream Cheese Frosting (p. 54)

No one will guess the secret ingredient in this delicious cake.

Preheat the oven to 350 degrees F.

Sift together all the dry ingredients, except the sugar. In a separate bowl, cream the butter until it is light yellow. Gradually add the sugar or honey. Stir the flour mixture and tomato puree into the butter and sugar alternately, adding a third of sauce or flour at a time. Stir until the batter is smooth. Fold in the nuts and raisins. Pour into a greased tube pan and bake for 45 minutes. When cooled, frost the cake with Cream Cheese Frosting.

Red and Green Tomato Treats

Soups/Salads
Side Dishes/Main Dishes

Red, White, and Green Soup

Yield: 10 servings
Time: 1 hour

2 tablespoons vegetable oil
3 tablespoons butter
4 very large onions (3 pounds), sliced in rings
5 medium-size red tomatoes, diced
4 green tomatoes, diced
⅓ cup unbleached flour
2 teaspoons dried mustard
1 cup dry red wine
6 cups beef or vegetable broth
2 teaspoons basil
2 teaspoons paprika
salt and pepper to taste
2 eggs, beaten
Parmesan cheese

In a large soup pot, heat the oil and the butter, and sauté the onion rings until tender and beginning to brown—about 15 minutes. Add the red and green tomatoes and continue cooking for 5 minutes. Stir in the flour and the mustard. Add the wine and stir slowly; then add the broth, basil, and paprika. Season with salt and pepper. Simmer the soup for 30 minutes.

Just before serving, add 1 cup of hot soup to the beaten eggs and mix well. Add the egg mixture to the soup pot and heat through, but do not boil or the soup will curdle. Serve garnished with grated Parmesan cheese.

When you mix egg yolks into a hot mixture it is important to warm them a bit before you add them or they may cook. That is why good cooks mix a small amount of warm sauce with egg yolks first and then add them to the pan.

October Lentil Soup

Yield: 8 servings
Time: 1½ hours

2 tablespoons vegetable oil
1 medium-size onion, diced
1 clove garlic, minced
2 stalks celery, diced
2 large carrots, diced
3 large green tomatoes, diced
1½ cups lentils
5 cups water
1 teaspoon thyme
½ teaspoon oregano
1 teaspoon basil
4 cups chopped red tomatoes, canned
   or fresh
3 tablespoons tomato paste (optional)
salt (or soy sauce) and pepper to taste
¼ cup dry sherry
1 cup sour cream

In a large soup pot, heat the oil, and sauté the onion, garlic, celery, carrots, and tomatoes for 10 minutes, stirring frequently. Add the remaining ingredients, except the sherry and sour cream, and bring to a boil. Cover and simmer for 1 hour, or until the lentils are tender. Put half of this mixture in a blender and process until smooth. Return the lentil puree to the soup pot. Add the sherry and cook for 10 minutes, uncovered. Correct the seasoning. Garnish with sour cream.

Mulligatawny Soup

Yield: 8 servings
Time: 1½ hours

2 tablespoons vegetable oil
1 medium-size onion, diced
1 medium-size carrot, diced
2 stalks celery, diced
1 cup turnips, diced
4 green tomatoes, diced
6 cups chicken stock
¼ cup tomato paste
2 teaspoons curry powder
dash cayenne pepper
½ cup unbleached flour
2 cups chicken, diced
2 cups cooked chick peas
salt and pepper to taste
½ cup shredded coconut

In a medium-size soup pot, heat the oil. Add the onion, carrots, celery, turnips, and green tomatoes. Sauté for 10 minutes, or until the vegetables are softened slightly. Add 5 cups of the chicken stock, the tomato paste, curry powder, and cayenne pepper.

In a small cup, mix the flour and 1 cup of chicken stock into a smooth paste. Stir the paste into the soup. Add the remaining ingredients, except the coconut, and simmer for 1 hour. Garnish with coconut and serve.

Mulligatawny soup is traditionally served as a smooth soup with pieces of chicken in it. I prefer my soups with chunky vegetable pieces. If you would like a smooth soup, puree the vegetables with a little broth after they have been sautéed and continue with the recipe.

Green Tomato Caponata

Yield: 8 servings
Time: 45 minutes, plus chilling time

½ cup olive oil
2 cups celery, diced
1 large onion, diced
6 cups unpeeled eggplant, cubed
4 green tomatoes, diced
½ cup wine vinegar
½ cup water
1 teaspoon honey
¼ cup tomato paste
2 tablespoons capers
½ cup parsley, minced
salt and pepper to taste

The secret of this dish is that each vegetable is sautéed separately to retain its flavor.

Serve caponata as an appetizer or salad, with Italian bread, or stuffed in tomatoes or pita bread.

In a large frying pan, heat 3 tablespoons of olive oil. Sauté the celery in the oil until just tender; do not brown. Remove the celery to a large bowl. Add 2 tablespoons of oil to the pan, and sauté the onion. Add the onion to the celery.

Heat the remaining olive oil, and sauté the eggplant for 5 minutes; then add the green tomatoes. Cook until soft. Remove the vegetables from the pan and add them to the celery and onion; leave the liquid in the pan. Add the vinegar, water, honey, tomato paste, and capers; simmer for 5 minutes.

Return the vegetables to the frying pan, add the parsley, and simmer for 10 minutes. Add salt and pepper and correct the seasonings. Cool this dish completely before serving.

Caponata keeps well in the refrigerator for several days.

Spanish Rice

Yield: 6–8 servings

Time: 30–45 minutes

2 cups red tomato juice or puree

2 cups water or chicken or
  vegetable broth

2 cups white or brown rice

2 tablespoons vegetable oil

1 medium onion, minced

1 stalk celery, minced

½ green pepper, minced

1 green tomato, minced

In a medium-size saucepan, heat the tomato puree and water to boiling, add the rice and cook until tender—15 minutes for white rice, 45 minutes for brown rice. In the meantime, heat the oil, and sauté the vegetables for 10 minutes. When the rice is cooked, stir in the vegetables and serve.

Enchilada Bake

Yield: 6 servings
Time: 1¼ hours

2 tablespoons vegetable oil
1 medium onion, diced
2 cloves garlic, minced
3 green tomatoes, diced
2 cups cooked kidney beans
salt and pepper to taste
3 tablespoons chili powder
12 corn tortillas
3 cups Red Taco Sauce (p. 91)
2 cups grated cheddar or Monterey Jack cheese

Preheat the oven to 350 degrees F.

In a medium-size saucepan, heat the oil, and sauté the onion, garlic, and green tomatoes for 10 minutes. Add the kidney beans, salt, pepper, and chili powder. Mash half of the beans with the back of a spoon. Cook for 10 more minutes, stirring occasionally.

In the meantime, grease a baking dish, and line it with the tortillas. Spread half of the bean mixture on top of the tortillas and cover with half of the Taco Sauce and half of the cheese. Continue layering, ending with a layer of cheese. Bake for 45 minutes. Cut in squares and serve.

Sweet and Sour Chicken

Yield: 6 servings

Time: 1½ hours

2 **tablespoons vegetable oil**

2 **tablespoons butter**

6 **pieces of chicken**

4 **medium-size green tomatoes, coarsely diced**

1 **carrot, sliced**

1 **green pepper, coarsely diced**

1 **onion, diced**

2½ **cups Sweet and Sour Sauce (p. 92)**

Preheat the oven to 350 degrees F.

In a large frying pan, heat the oil and the butter, and brown the chicken on both sides. Transfer the chicken to a baking dish and arrange the vegetables around the chicken. Pour the Sweet and Sour Sauce over the top, and bake for 45–60 minutes.

Stuffed Meat Loaf

Yield: 6–8 servings
Time: 1¾ hours

Meat loaf:
1½ pounds ground beef
2 eggs
1 medium onion, finely diced
2 cloves garlic, crushed
¾ cup bread crumbs
½ cup tomato sauce, puree, or catsup
2 tablespoons Worcestershire sauce
1 teaspoon salt

Stuffing:
2 tablespoons vegetable oil
1 small onion, diced
2 large green tomatoes, diced
2 stalks celery, diced
2 carrots, diced
1 cup peas, fresh or frozen
1½ teaspoons basil
1 teaspoon thyme
½ cup Parmesan cheese
½ cup bread crumbs
salt and pepper to taste

This is best when served with the Celery Tomato Sauce (p. 94).

Preheat the oven to 350 degrees F.

In a large mixing bowl, mix together all the ingredients for the meat loaf. Spread the mixture onto a piece of waxed paper in the shape of a rectangle ½ inch thick.

Heat the oil, and sauté the onion, tomatoes, celery, and carrots for 10 minutes. Add the remaining ingredients. Spread this mixture over the meat loaf rectangle.

Using the waxed paper, roll the meat loaf in jelly-roll fashion. Place the meat loaf, seam side down, on a baking sheet. Bake for 1 hour.

Simmering Chicken Pot

Yield: 6 servings

Time: 1½ hours

2 tablespoons butter
6 pieces of chicken
2 cloves garlic, minced
1 medium onion, cut in rings
1 green pepper, chopped
½ pound mushrooms, sliced
6 green tomatoes, chopped in large pieces
2 red tomatoes, chopped in large pieces
1½ cups chicken broth
½ cup white wine (optional)
2 teaspoons thyme
1 teaspoon marjoram
salt or tamari sauce and pepper to taste
¾ cup unbleached flour

This is delicious served over Yellow Confetti Rice (p. 44).

In a Dutch oven, melt the butter and brown the chicken on both sides, until just golden. Remove the chicken from the pot. Sauté the garlic, onion, peppers, and mushrooms for 5 minutes. Add the remaining ingredients, except the flour. Cook for 10 minutes. Return the chicken pieces to the pot. Ladle sauce on each piece. Cover and simmer for 1 hour.

Twenty minutes before you are ready to eat, remove the chicken from the pot and keep it warm. Put the flour in a small mixing bowl and ladle 1 cup of hot broth from the pot into the flour. Stir briskly to remove the lumps. Add the flour mixture to the pot and stir, being careful to get rid of all the lumps. Return the chicken to the pot and cover. Simmer the mixture gently until you are ready to serve.

159 Terrific Tomato Ideas

What's for Dinner Quick!

If you're a cook who likes helpful hints for menu planning, but really enjoys a free rein to create your meals, this section is for you. Here are some food combination ideas to help you plan easy, tomato-based side dishes, sauces, and casseroles, by combining, mixing, and matching certain foods and herbs.

For side dishes, sauté tomatoes with any one or a combination of vegetables, and flavor with an herb or combination of herbs, such as the ones suggested on p. 134.

To make a special sauce, combine sautéed vegetables with meats or fish, add some flavoring liquids, and serve over a grain or pasta.

To make a tomato-based casserole, layer the sauce with the grains or pasta and finish with a cheese and crumb topping. Bake for 30–60 minutes.

Vegetables to Combine with Tomatoes

Green cabbage
Corn
Green peppers
Mushrooms
Eggplant
Zucchini
Broccoli
Cauliflower

Onions
Peas
Lima beans
Chick peas
Black olives
Celery
Okra

Cooked Meats and Fish to Mix In

Chicken
Turkey
Ham
Bacon
Ground beef
Lamb, ground or cubed
Beef, cubed
Pork, ground or cubed
Sausage, sweet or hot

Shrimp
Filets of sole, bass,
    or turbot
Clams
Tuna
Salmon
Red snapper
Mussels

Liquids for Flavor

Wine
Vegetable broth
Chicken broth
Beef broth
Prepared mustard
Tomato sauce
Sour cream
Cream

Cheeses to Match

Parmesan
Cheddar
Swiss
Mozzarella
Blue cheese
Feta

Serve On

Pasta of any shape
Cous cous
Brown rice
White rice
Polenta (cornmeal mush)
Spaghetti squash
Cooked leaf spinach
Toast

And Top With

Sesame seeds
Bread crumbs
Wheat germ
Cracker crumbs

Salad Combos

Any of these salad combinations are great with sliced, diced, or quartered tomatoes. Slice, dice, or chop the vegetables. Mix with the tomatoes and top with a simple vinaigrette dressing.

Mexican

Avocado
Green pepper
Red pepper
Chick peas
Cooked corn

Farm-style

Carrots
Green pepper
Kohlrabi
Cucumbers

French

Mushrooms
Parsley
Fresh peas

Oriental

Broccoli (parboiled)
Water chestnuts
Cashews
Soy sauce

Greek

Black olives
Red onions
Green peppers
Feta cheese
Dill

Herb Combinations

Whether you are throwing together a quick casserole or creating a new tomato dish from scratch, you can use these herb combinations to achieve special, authentic tastes.

Mexican

Chili powder
Cumin
Garlic
Onion
Hot peppers
Coriander leaves
    or cilantro

Southern French

Onion
Garlic
Thyme
Basil
Bay leaf

French Canadian

Cinnamon
Sugar
Nutmeg

Indian

Curry powder
Hot peppers
Cumin
Turmeric

Oriental

Soy sauce
5-Spice-Powder
    or anise
    or fennel
    or ginger

Greek

Lemon
Garlic
Parsley
Dill
Thyme

Northern French

Tarragon
Parsley
Onion

Italian

Onion
Garlic
Basil
Thyme
Oregano
Parsley
Marjoram

English

Sage
Celery seed
    or mustard

Russian

Caraway seeds
Dill

Seafood

Horseradish
Lemon
Tabasco

The Art of Tomato Garnishes

Sometimes a special meal calls for some special touches. You can add a sprig of parsley to a serving platter, or sprinkle some chopped scallions on top of the dish. But tomatoes add the most distinctive, gay touches to a meal. Sliced or wedged tomatoes encircling a platter add lovely color. Stuffed tomatoes are another great garnish for a serving platter, especially if you take the time to make a tomato crown or basket.

Here are some of the tomato garnishes I use to transform ordinary meals into special "catered affairs."

Tomato Roses

The tomato rose is the supreme tomato garnish. It looks so realistic, it can compete with a rose straight out of the garden. I like to put a large rose in the center of a dish with parsley as its leaves. You can also use a large rose and a small rose together, cutting green peppers as leaves or setting the roses on a bed of watercress.

Tomato roses are not hard to make, but don't expect perfect results the first time. Increase your chances for success by sharpening a knife for the job, and always select firm, ripe tomatoes. The tomatoes should be medium in size.

A tomato rose is made from a continuous strip of tomato peel, cut about ¾ inch wide. Experience will teach you how thick to make this strip. It should be thick enough so that it does not break, but not so thick that it won't bend. Ready to start?

1. Working from the stem end, insert the point of your tomato knife into the tomato to the depth you want for the peel. Slide the knife so that the edge of the blade is parallel to the surface of the skin.

2. Begin peeling around the tomato in a continuous strip. Be sure to hold the tomato near the table so the strip does not dangle and break. (If your strip breaks, don't despair, you can still use it; more on that later.)

How to make a Tomato Rose

3. At the bottom of the tomato, the strip will end with a little curl that becomes the "hook" to hold your rose together. You are now ready to wrap your rose.

4. Take the beginning of the tomato strip and wrap it around your index finger. Continue wrapping the strip on top of itself until you reach the end.

5. Hook the tail into the previous ring, or petal. Slip the rose off your finger, and, if necessary, tighten the center of the rose by turning the center of the spiral.

6. Invert the rose and position it on the serving platter.

If you have a few broken pieces, try this method. Take the longest piece and start rolling it with one edge against the table top so you can see the top of the rose forming as you roll. Before you come to the end of the first piece, work in the second piece, and so on, until you have worked the last piece in. Hook in the end of the last strip into the rose and move the rose carefully to the serving platter.

The roses can be stored in the refrigerator on a plate covered with a damp towel for a few hours.

Tomato Peony

Here's another tomato garnish that closely resembles a flower, and it is easier to make than the rose. The peony is a rather large garnish and should be used on a large serving platter.

Hold a firm tomato stem side down. Starting on one side, make a cut across the top of the tomato three-quarters of the way through the tomato. Make a second cut, parallel to the first, about ¼ inch beyond. Continue making parallel cuts, each three-quarters of the way through, across the tomato, leaving ¼ inch between cuts. Turn the tomato a quarter turn, and make a second series of deep cuts ¼-inch apart across the tomato to create little squares. Gently spread the petals, being careful not to break any.

Cut 2 or 3 leaf shapes from a green pepper and place them around the blossom.

Butterfly Garnish

This garnish will delight kids as well as adults. It is one of the best garnishes you can make with green tomatoes.

Take a large firm red or green tomato, and cut it into ¼-inch slices. Cut each slice in half and dip the cut edges into 2 tablespoons of minced parsley to coat them. Place the slices back to back on the dish you are decorating. Cut 2 small pieces of chives or scallion greens, place them on the tomato slices to make eyes. Arrange 2 longer pieces of chives to make antennae.

Tomato Accordian

Eggs, cucumbers, or peppers can be used with this garnish, depending on what colors you wish to emphasize with your garnish.

Remove a thin slice from the stem end of the tomato to make sure the tomato stands flat. Set the tomato on the cut end. Make 5 deep slits into the tomato. Slip slices of hard-boiled egg, cucumber, or pepper between the slits.

Mushroom Bowls

This garnish can be used with hot or cold foods. The mushrooms can be raw or lightly sautéed.

Remove the stem from the mushroom. Set the mushroom cut side up on a platter. Place a half cherry tomato inside the mushroom where the stem had been. Set 3 mushrooms together on a bed of parsley for a very unusual garnish on the side of your platter.

With a cold dish, you may want to garnish with zucchini or cucumber slices and halved cherry tomatoes, instead of mushrooms.

Casserole Garnish

Cut a tomato into wedges. Arrange the wedges in an overlapping circle at the center of any casserole dish during the last 20 minutes of baking time.

tomato peony

tomato rose

tomato butterfly

mushroom bowls

cordian

casserole garnish

tomato bowl

basket

crown

tomato
wheel

tomato
blossom

stuffed tomato slice

striped
tomato

There's More Than 1 Way To Stuff a Tomato

The ordinary garden tomato is transformed into something quite exotic when filled with a delicious savory filling. Everything from gourmet treats to leftovers can be presented in the natural serving bowl that a hollowed-out tomato becomes. Little stuffed cherry tomatoes make colorful, easily served hors d'oeuvres or appetizers. Full-sized tomatoes can be stuffed with salads or raw vegetables for a first course or light lunch, or they can be stuffed with meats and cooked vegetables as a main course.

6 Tomato Shapes

I love to hear the excited comments when I serve a stuffed tomato in a unique style. People always say "I never knew you could do *that* to a simple tomato." Tomato crowns, tomato wheels, tomato baskets convert simple meals into rare feasts for the eye.

Be sure that you work with a good sharp paring knife. Since the stem end of the tomato is flatter, I usually make that end the bottom and then the tomatoes don't roll off the plate. This is particularly important with cherry tomatoes. You can bake stuffed tomatoes in muffin cups to help them hold their shape and not roll over. All stuffed tomatoes can be peeled first to make them easier to eat, but I prefer my tomatoes with skins because they are more nutritious that way.

Save the insides of the tomatoes! The pulp can be used in soups, stews, gravies, and salad dressings.

The Basic Shape

For a basic stuffed tomato, simply cut off the bottom of the tomato, scoop out the insides, drain, and stuff. But there are many variations to that theme.

Tomato Baskets

A whimsical change for stuffed tomatoes is the Tomato Basket. It is served filled with cold foods. Use only large firm tomatoes, and make

How to make a Tomato Basket

sure your knife is extra sharp for this operation. I like to use a serrated knife.

1. Looking down at the top of the tomato (the stem end), slice into the tomato just right of the stem scar. Draw your knife halfway down into the tomato, stop, and bring the knife back up.

Make another cut, parallel to the first on the other side of the stem scar, about ¾ inch from the first slice.

2. Now make 2 more cuts into the sides of the tomatoes so that these cuts meet the slices you made in step 1.

3. Remove the wedges of tomato you just cut. What remains is the basket handle and basket.

4. Using your knife, remove the pulp and seeds from the handle. Leave as much of the flesh of the tomato as possible, so that the handle stays firm.

5. Remove as much seed and pulp as possible from the basket.

6. Stuff the basket with a cold salad. If you like, add a sprig of parsley as a "ribbon" decoration.

Tomato Bowls

Use a tomato bowl instead of a china bowl to serve relishes. A hollowed-out tomato can hold tiny pickled onions, olives, relishes, or sliced pickles on a sandwich plate or cold cut plate for a picnic or buffet.

Select a large, firm tomato. Remove the bottom of the tomato. Scoop out the insides, and allow the tomato to drain for a few minutes on a paper towel. Set the tomato on a platter, stem side down, and fill.

Tomato Crowns

Here's a classic stuffed tomato shape that is good for hot and cold foods. Use a firm tomato of any size. The crown is made by a series of V-shaped cuts all the way around the tomato.

To start, hold the tomato in one hand and push your small, sharp paring knife into the tomato at a 45-degree angle, midway down the side of the tomato. Remove the knife. Now make another cut into the tomato at a 45-degree angle in the opposite direction. The bottom edge of the second cut should meet the bottom edge of the first cut so that a V is formed. Continue making these V-shaped cuts around the entire tomato. Pull the 2

halves apart, and scoop out the seeds. Fill each half with the stuffing of your choice. Or fill the first half with your stuffing, and top with the second half.

To use a tomato crown as a garnish, fill the cups with sprigs of parsley and set the crowns on a serving platter filled with other foods.

Tomato Wheels

Tomato wheels are filled with cold foods. They are easy to make.

Holding the tomato with the stem side down, make a long diagonal cut from the near top of the tomato to halfway down its side. From the bottom edge of this cut, make a second cut up, in the opposite direction, to make a long V-shaped cut. Make a second V-shaped cut to match the first. Continue until you have made 8 identical wedges (16 cuts). Remove the top of the tomato and the wedges you have made. Scoop out a little of the center. This section can be filled with a cold stuffing, such as ham salad. Now take thin slices of cheese. Using a 1–1½-inch fluted cookie cutter, make 8 rounds of cheese. Put a cheese circle in each wedge.

Stuffed Tomato Blossom

You can use either peeled or unpeeled toma-toes for Tomato Blossoms. Serve hot or cold foods in the blossom.

Turn the tomato stem end down, and cut the tomato into 7 petals that extend ⅔ of the way down the sides. Gently spread the petals out. Scoop out the flesh of the tomato, and fill the center with your filling. Then close the petals up over the stuffing.

If you are planning to use a very liquid filling, cut the petals only ⅓ of the way down the tomato.

Striped Tomato

This tomato should be filled with a firm, cold salad. It makes a wonderful low-calorie alternative to the sandwich.

Cut each tomato into 3 thick slices and neatly spread your filling between the layers. Then carefully restack the layers.

Stuffed Tomato Slices

Here is another low-calorie alternative to bread for serving with a cold salad.

Cut a large tomato into thick slices. Top each slice with a slightly smaller ring of green pepper. Spoon your filling into the green pepper ring and garnish.

Bed Down That Tomato!

For the best effect, serve your stuffed tomatoes on a beautiful dish. Make a bed for the tomatoes by arranging fresh, crisp lettuce leaves or shredded vegetables on the serving dish. Then set your tomatoes on top. Here are some suggestions.

Lettuce	Watercress	Sprouts	Shredded carrots
Spinach	Parsley	Shredded cabbage	Swiss chard

Be adventurous when you decorate your stuffed tomatoes. Make a green trim on the edge of your tomatoes by dipping the tomato in finely chopped fresh parsley. Or choose one of the following garnishes to sprinkle on top.

Herbs: Parsley, basil, oregano, mint, sage, dill, or chives.

Seeds: Sesame, pumpkin, or poppy.

Nuts: Walnuts, almonds, cashews, or hazelnuts.

Vegetables: Grated or diced scallions, radishes, carrots, green peppers, olives, pickles, capers, or beets.

Stuffed Tomato Recipes

Now that you are all set with your tomato bowls and baskets, here are some ideas for filling them. On the pages that follow are some recipes I use. Or make up your own and let your imagination be your guide.

Egg Salad Stuffings

Dress up your usual egg salad with any one of these ingredients to make a fancy luncheon dish. Stuff into tomato shells and serve on a bed of lettuce.

Chinese Choice: 2 tablespoons water chestnuts, minced.

Spring Salad: ¼ cup celery, minced with 1 teaspoon celery seed.

Garden Side: 2 tablespoons minced parsley.

Picklers' Preference: 1 tablespoon dill. Garnish with cucumber slices.

Bombay-style: 2 teaspoons curry powder.

Genoa Eggs: ¼ cup Parmesan cheese and 1 teaspoon basil.

The Parisians: 1 tablespoon capers, minced.

Polish Approach: 1½ teaspoons caraway seeds and 1 tablespoon minced pickles.

Deep Sea Diver: 2 teaspoons anchovy paste, garnished with anchovy wheels (rolled up anchovy filets).

Western Range: 1 scallion, finely minced, and ¼ cup finely minced ham.

Chicken Salad Stuffings

Take your favorite chicken salad recipe, jazz it up with one of these variations, and stuff it into tomato shells for an attractive light meal.

Avocado Curry: Dice an avocado and add it to the chicken salad, with 2 teaspoons curry powder, a squeeze of lemon juice, and mayonnaise.

Chicken Waldorf: Add diced apple, green grapes, walnuts, and mayonnaise.

Sunshine Salad: Add 1–2 tablespoons honey, 1 orange, divided into sections, 2 teaspoons poppy seeds, and mayonnaise.

Cheese Stuffed Tomatoes

Make a special cheese spread to stuff into a tomato cup.

Farm-style: Combine cottage cheese with basil, chives, dill, and scallions.

Liptauer: A traditional Austrian spread made by combining 1 cup cottage cheese, 2 tablespoons paprika, 1½ teaspoons caraway seeds, 1 scallion, minced, and 2–3 tablespoons Dijon-style mustard. Mix in a blender or processor until smooth.

Blue Cheese: Combine cottage cheese with blue cheese and onion.

Greek-style: Combine cream cheese with feta cheese, parsley, and dill.

Coastal Delight: Combine cream cheese with smoked oysters, lemon juice, and Tabasco sauce.

The Hero: Combine diced cheddar or Swiss cheese with raw diced vegetables and slices of pepperoni. Pour a vinaigrette dressing over the filling.

Hot Vegetable Cups

Tomatoes make natural bowls or containers for garden vegetables. Hot vegetables should be parboiled (for 2–3 minutes) and then reheated in the tomato cups.

Peas: Fill the cups and bake for 10 minutes.

Broccoli: Parboil the broccoli florets. Place the florets, stem first, in the cups. Dot with butter and bake for 5–10 minutes.

Cauliflower: Parboil cauliflower pieces. Place in tomato cups. Melt 2 tablespoons butter and brown ½ cup bread crumbs. Sprinkle the crumbs over the cauliflower and bake for 10 minutes.

Green Beans: Parboil the beans. Decorate with almonds and bake for 10 minutes.

Corn: Fill the tomato cups with cooked whole kernel corn. Garnish with minced parsley.

Mushrooms: Sauté mushrooms and diced onions in butter. Reheat in tomato cups.

Hummus Stuffed Tomatoes

2 cups cooked chick peas
2 cloves garlic, minced
3 heaping tablespoons tahini
  (sesame seed paste)
¼ cup lemon juice
salt to taste
50 red cherry tomatoes, with insides
  scooped out

This filling is wonderful when served as an appetizer in cherry tomato cups.

Mix all ingredients, except the tomatoes, in a blender (add a little water if necessary) or in a food processor. Blend until smooth. Stuff into tomatoes. Serve the remaining Hummus with vegetable sticks or Syrian pita bread.

Tofu Tahini Salad Cups

3 tablespoons tamari (soy sauce)
3 tablespoons water
1 garlic clove, minced
½ pound tofu, cut in small cubes
⅓ cup tahini (sesame seed paste)
2 tablespoons lemon juice
4 tomatoes, with insides scooped out

Mix the tamari, water, and garlic together in a bowl. Add the tofu and let stand for 15 minutes. Remove the tofu. Add the tahini and lemon juice to the remaining liquid and mix well. Stir this liquid into the tofu. Stuff into tomato cups and serve.

Mushroom Nut Paté Cups

¼ cup butter
¾ pound mushrooms, sliced
1 small onion, minced
1 clove garlic, minced
1 cup pecans
½ teaspoon thyme
dash nutmeg
salt and pepper to taste
approximately 50 red cherry tomatoes

In a medium-size frying pan, melt the butter and sauté the mushrooms, onion, and garlic until they are soft. Combine all the ingredients, except the tomatoes, in a blender or food processor and blend until smooth. Chill for ½ hour.

Just before serving, scoop out the insides of the cherry tomatoes and stuff the tomatoes with paté. Serve the remaining paté with crackers or bread.

Tofu Spread

½ pound tofu, cut in chunks
2 scallions, diced
¼ cup tahini (sesame seed paste)
2 tablespoons tamari or soy sauce
2 tablespoons parsley
50 cherry tomatoes, with insides
   scooped out

Put all ingredients in a blender or food processor and blend until smooth. Stuff into red cherry tomato cups. Serve the remaining spread on crackers or celery sticks.

Rosy Snow-Capped Tomatoes

12 small to medium-size, ripe red
   tomatoes
1 carrot, grated
1 green pepper, finely diced
1 small onion, minced
½ cucumber, grated
1 cup cottage cheese
4 ounces cream cheese
1½ cups Rosy Cream Dressing (p. 83)
salt and pepper to taste
1 cup heavy cream, whipped

Served slightly frozen, this unusual stuffed tomato dish is perfect for a blistering August day.

Hollow out the tomatoes and save the insides for the Rosy Cream Dressing. Turn the tomatoes upside down on a paper towel to drain.

Mix the remaining vegetables with the cottage cheese, the cream cheese, ½ cup of Rosy Cream Dressing, and salt and pepper. Stuff this mixture into the tomatoes, and turn them upside down in a shallow casserole pan or freezer pan.

Fold the remaining Rosy Cream Dressing into the whipped cream. Pour this dressing over the tomatoes, cover lightly, and place them in the freezer for 2 hours, or until the dressing begins to freeze slightly. Serve.

Chicken Coronation in Tomatoes

8 firm, medium-size red tomatoes
2 tablespoons vegetable oil
1 medium onion, diced
1 large green pepper, diced
1 heaping teaspoon curry powder
¾ cup red wine
½ cup water
½ cup catsup
1 tablespoon apricot jam
6 cups diced cooked chicken
½ cup mayonnaise
¾ cup heavy cream, whipped
salt and pepper to taste
lemon juice to taste

England is not always remembered best for its food, but some English dishes just shouldn't be missed. This is one of them. Chicken Coronation was invented for the celebration of Queen Elizabeth II's coronation, and it is fit for a queen.

Cut off the bottoms of the tomatoes. Scoop out the pulp. Turn upside down on paper towels to drain.

In a large frying pan, heat the oil, and sauté the onion and pepper for 5 minutes. Add the curry powder, and cook for 5 minutes. Add the wine, water, catsup, and apricot jam. Simmer gently for 10 minutes, or until the sauce is somewhat reduced. Set the sauce aside to cool.

Mix the cooled sauce with the chicken. Fold in the mayonnaise and whipped cream. Correct the seasoning with salt and pepper and lemon juice.

Fill the tomato cups with the chicken and serve.

Spinach Cream in Tomato Cups

6 firm, medium-size tomatoes
2 tablespoons butter
2 cloves garlic, minced
2 tablespoons unbleached flour
¾ cup milk or cream
2 cups chopped cooked spinach
½ teaspoon thyme
2 teaspoons lemon juice
salt and pepper to taste

Preheat the oven to 350 degrees F.

Cut the tops off the tomatoes and scoop out the insides. Turn the tomatoes upside down on a paper towel to drain.

In a medium saucepan, melt the butter, and sauté the garlic for 5 minutes; do not brown. Stir in the flour. Slowly add the milk, stirring constantly to remove any lumps. Add the remaining ingredients, and correct the seasoning.

Pour the filling into the tomato cups. Place the tomatoes on a baking dish. Bake for 20 minutes and serve.

Italian Spinach Tomatoes

6 firm, medium-size red tomatoes
2 cups chopped cooked spinach
1½ cups Parmesan cheese
½ teaspoon garlic powder
1 teaspoon thyme
1 teaspoon basil
1 egg
salt and pepper to taste

Preheat the oven to 350 degrees F.

Cut the tops off the tomatoes and scoop out the insides. Turn the tomatoes upside down on a paper towel to drain.

In a separate bowl, combine the remaining ingredients, saving ¼ cup Parmesan cheese.

Divide the spinach mixture among the tomatoes. Sprinkle the top of each tomato with Parmesan cheese. Bake for 15 minutes and serve.

Stuffed Spanish Tomatoes

8 large, firm red tomatoes
1 pound cooked bulk sausage
1 large onion, diced
½ pound mushrooms, sliced
1½ cups ham, finely diced
1 cup bread crumbs
1 cup Parmesan cheese
salt and pepper to taste
2 cloves garlic, minced
2 teaspoons basil
2 tablespoons unbleached flour

Preheat the oven to 350 degrees F.

Hollow out the inside of the tomatoes and reserve the pulp for the sauce. Turn the tomato shells upside down on a paper towel to drain.

Brown the sausage, remove the sausage from the pan, and set it aside. Drain off all but 2 tablespoons of the fat. In the remaining fat, sauté half of the diced onion and all of the mushrooms until the onion is translucent and the mushrooms are soft. Combine the onion and mushrooms with the sausage. Add the diced ham, half the bread crumbs, and half the Parmesan cheese. Add salt and pepper to taste.

Divide the stuffing among the tomatoes. Sprinkle the remaining bread crumbs and Parmesan cheese on top.

To make the sauce, combine the pulp from the tomatoes, the remaining diced onion, the garlic, and the basil in a saucepan. Simmer for 15 minutes. Pour this mixture into a blender, add the flour, and blend until smooth. Add salt and pepper to taste. Pour the sauce over the tomatoes and bake for 30 minutes or until the sauce is bubbling hot.

Greek Stuffed Tomatoes

6 medium-size red tomatoes
2 tablespoons olive oil
1 small onion, diced
2 cups chopped cooked spinach
2 teaspoons basil
½ cup bread crumbs
1 egg
½ pound feta cheese, crumbled
salt and pepper to taste

Preheat the oven to 350 degrees F.

Cut the bottoms off the tomatoes and scoop out the insides. Put the tomatoes upside down on a paper towel to drain.

In a small saucepan, heat the oil, and sauté the onion for 10 minutes, until translucent. Combine all the remaining ingredients and add to the onions.

Stuff the filling into the tomatoes. Bake for 15 minutes and serve.

Best Brunch Eggs

½ cup butter or margarine
½ cup unbleached flour
2 cups milk
1 pound cheddar cheese, grated
1 tablespoon Dijon-style mustard
dash Worcestershire sauce
salt and pepper to taste
6 large, red tomatoes, with insides
   scooped out
salt and pepper to taste
1 cup grated cheddar cheese
6 eggs

Melt the butter in a saucepan over medium heat. Stir in the flour, making sure there are no lumps. Slowly add the milk, a little at a time, stirring after each addition to remove all the lumps. Add the pound of cheese, the mustard, Worcestershire sauce, and salt and pepper to taste. Continue cooking, stirring constantly, until the sauce is thick and smooth. Set aside.

Preheat the oven to 350 degrees F.

Grease a baking dish, and arrange the tomatoes in the dish. Sprinkle the insides of the tomatoes with salt, pepper, and the remaining cheese. Break the eggs into the tomatoes, and spoon the cheese sauce on top. Bake for 20–30 minutes, or until the eggs are set.

The best way to prevent lumps in your white sauce is to heat the milk and add it slowly to the roux (flour and butter paste).

Index